Geraldo Henrique Vieira

Physical treatments in post-harvest of 'Thompson' Atemoia

Geraldo Henrique Vieira

Physical treatments in post-harvest of 'Thompson' Atemoia

Use of organic acids, ionizing irradiation and active modified atmosphere

ScienciaScripts

Imprint
Any brand names and product names mentioned in this book are subject to trademark, brand or patent protection and are trademarks or registered trademarks of their respective holders. The use of brand names, product names, common names, trade names, product descriptions etc. even without a particular marking in this work is in no way to be construed to mean that such names may be regarded as unrestricted in respect of trademark and brand protection legislation and could thus be used by anyone.

Cover image: www.ingimage.com

This book is a translation from the original published under ISBN 978-613-9-68838-8.

Publisher:
Sciencia Scripts
is a trademark of
Dodo Books Indian Ocean Ltd. and OmniScriptum S.R.L publishing group

120 High Road, East Finchley, London, N2 9ED, United Kingdom
Str. Armeneasca 28/1, office 1, Chisinau MD-2012, Republic of Moldova, Europe
Printed at: see last page
ISBN: 978-620-8-23057-9

SUMMARY

I dedicate this work to my beloved parents, Geraldo Martins Vieira and Edésia da *Conceiçao ^Marques* Vieira and *my wife Iza Mara Lopes Vieira.*

ACKNOWLEDGMENTS

To God.

To my parents for the training they gave me and for their trust and support.

To my wife, for her support and welcome in difficult times.

To my advisor Prof. Dr. Rogério Lopes Vieites, for his teachings and patience.

To the Department of Horticulture, through its collaborators, whom I affectionately call "guardian angels"; and especially the professionals Karina Aparecida Furlaneto and Flâvia Aparecida de Carvalho Mariano Nasser.

To the Postgraduate Program in Agronomy (Horticulture) at UNESP, Câmpus Botucatu S.P., which, together with the INOVA Sao Paulo program, provided this opportunity.

To IPEN (Instituto de Pesquisa Energéticas e Nucleares), which, through the Centro de Tecnologia das Radiaçôes - CTR IPEN/CNEN - SP, was able to provide the technical support needed to apply ionizing radiation to one of the experiments.

And to everyone who, directly or indirectly, helped make this doctoral thesis a reality.

"We are not what we know, we are what we are willing to learn."

Concil

SUMMARY

The aim of this work *was* to study the 'Thompson' cultivar of atemoya under three different post-harvest treatments: the use of organic acids, active modified atmosphere and ionizing irradiation (^{60} Co). The fruit was selected and stored in a cold room under conditions of 15 ± 0.2°C and 90 ± 2% RH. For each experiment, there were three replications and two fruits per replication. Six treatments were carried out: two organic acids (ascorbic and citric); at concentrations of 1%, 2% and 3% each treatment was submerged in its organic acid solution for 10 minutes and then dried in the shade. There were four treatments in the active modified atmosphere: T0-21% O_2, 0.03% CO_2 and 78% N2 (ambient); T1 -4% O_2, 5% CO_2 and 91% N_2, T2 -4% O_2, 6% CO_2 and 90% N2, T3 -4% O_2, 7% CO_2 and 89% N2 and T4- 4% O_2, 8% CO_2 and 88% N2. In the ionizing irradiation experiment (^{60} Co), the following treatments were used: T0= No irradiation (control), T1=0.2kGy, T2=0.4kGy, T3=0.6kGy, T4=0.8kGy, T5=1.0kGy and T6= 1.2kGy. In all the comparisons of all the experiments, the samples were analyzed every three days up to eighteen days of storage. In the control treatment, the fruit was washed under running water and then immersed in a chlorine solution of 150 mg L^{-1} for 10 minutes and then dried at room temperature, with the exception of the ionizing irradiation control treatment, which did not receive this treatment. The following analyses were carried out on all treatments: peel and pulp chromatology, loss of fresh mass, soluble solids (SS), titratable acidity (TA), pH, ascorbic acid (AA), reducing sugars, respiration rate, specific polyphenoloxidase (PPO) and peroxidase (POD) activity, total antioxidant activity and total phenolic compounds. The experiments were conducted in a completely randomized design (D.I.C.), in a factorial scheme (treatment x storage), consisting of the treatments from the different experiments and eight storage times. The results were subjected to analysis of variance and the means were compared using the Tukey test at a 5% probability level and regression was performed for the analyses during the storage period. The results showed peak respiration in the atemoya on the sixth day for citric acid at 1% and 3% and for the active modified atmosphere with 4% O_2 and 8% CO_2 on the ninth day for the 1.2kGy dose.

Keywords: *Annona* x *atemoya* Mabb., enzymatic activity, ionizing irradiation, organic acids, active modified atmosphere.

GENERAL INTRODUCTION

The atemoia belongs to the anonaceae family, which has 128 genera and 2,300 different species; its diversity in Brazil is characterized by 29 genera, with approximately 385 species, of which 128 species are endemic (JUDD, 2009; MARTINELLI, 2013).

The atemoya is a hybrid, resulting from the crossing of a species originating from *Anonna cherimola* Mill. (cherimoya), which occurs spontaneously in the Andean regions of Chile, Peru, Bolivia, Ecuador and in places with a mild climate and *A. squamosa* L. (conde fruit, ata or pine cone) originating from the lowlands of Central America; this hybridization characterizes the *atemoya* as *Annona* x *atemoya* Mabb.

The Count's fruit is so named because it was introduced to Brazil by the Count of Miranda in 1626 in the state of Bahia; its fruit was analyzed by Maia et al. (1986) and Rego et al (1989), showed 54% pulp, 39% skin and 7% seeds; its pulp had soluble solids values of 5.24 °Brix, pH 4.3 and 0.88% total titratable acidity; according to Manica (1997), the cherimoya fruit has a small number of seeds (21 to 41 per fruit), with 54 to 71% pulp and soluble solids levels between 15 and 30 °Brix.

The hybridization that gave rise to the atemoia occurred in the USA in the state of Florida in 1908. The first work on the atemoia took place in the state of São Paulo in the 1950s at the Campinas Agronomic Institute (IAC), with the production of seedlings at the Seedling Production Center in São Bento do Sapucai; The main varieties found in Brazil are Thompson and Gefner, which according to Neves (2009) have pulp percentages ranging from 45.2% for 'Thompson' to 61.8% for 'Gefner', with soluble solids contents of 26.1 °Brix and 22.8 °Brix respectively.

Brazil's production of fresh fruit in 2016 was 39,305,097 tons in a harvested area of 1,991,445 hectares and generates an average of three jobs per hectare. The domestic market consumes around 97% of the total fresh fruit produced in the country and per capita consumption is 57 kilos (ANUARIO BRASILEIRO DE FRUTICULTURA 2017).

According to the Brazilian Program for the Modernization of the Horticultural Market (PROHORT), CEAGESP (Companhia de Entrepostos e Armazéns Gerais Unidade da Grande Sâo Paulo) was responsible for selling 2,970,597 kg of atemoya fruit in 2016, which corresponds to 81.6% of the total volume sold in that year (CONAB, 2017).

Post-harvest losses in fruit in Brazil amount to 30% or more of production. These losses are linked to all items in the production chain, such as transportation, packaging, storage and processing. In addition to these factors, others can occur due to indirect actions such as the incidence of pathogens, darkening, loss of color (saturation, tone and luminosity), texture, smell and other factors that can be decisive for the consumer in their choice.

The atemoia is a climacteric fruit, which characteristically maintains its ripening process after harvest under conditions of temperatures between 10° and 25°C, and at a relative humidity of between 85 and 95% (CHITARRA; CHITARRA, 2005).

Storage at low temperatures, immediately after harvest, is the most widely used technique for prolonging fruit preservation; it influences respiratory activity and leads to lower losses in terms of aroma, flavor, texture and color, among other fruit quality attributes (BRON et al., 2002; LICHTER et al., 2000), 2000); a fact which justifies the storage of the atemoyas in a cold room under conditions of 15 ± 0.2°C in this experiment; this is coupled with the use of methodologies such as the active modified atmosphere which, together with the use of refrigeration, can delay the ripening of the fruit; because adequate levels of O_2, CO_2 and N2 can reduce the intensity of metabolic processes which interfere with the increase in post-harvest useful life (CHITARRA; CHITARRA, 2005); another example can be found in VIEITES (2009), who cites the use of ionizing radiation as a method of improving fruit preservation, with the possibility of reducing the incidence of pathogens which may influence the natural ripening process, and this technique is considered safe.

Other methods can also be used to preserve fruit, such as the use of organic acids to reduce enzymatic browning. The acids act directly to reduce the pH of the fruit, thus reducing the activity of the enzymes involved in the metabolic processes that cause browning, such as polyphenol oxidase and peroxidase.

Brazil is the third largest fruit producer in the world, behind only China and India, but it has high rates of post-harvest losses, which motivated the development of this work using organic acids (ascorbic acid and citric acid), ionizing irradiation^{60} CO and active modified atmosphere.

CHAPTER 1

Use of organic acids in post-harvest on the 'Thompson' atemoya cultivar

SUMMARY

The experiment with the use of organic acids on 'Thompson' cv. atemoia took place in the post-harvest laboratories of the Faculty of Agronomic Sciences - FCA - "Júlio de Mesquita Filho" UNESP - Campus of Botucatu S.P., with the aim of evaluating their action in maintaining post-harvest color. For each treatment, there were three replicates with two fruits per replicate. Six treatments were carried out, with two organic acids (ascorbic and citric); at concentrations of 1%, 2% and 3% each; the atemoias were submerged in the organic acid solution for 10 minutes and then dried in the shade. In the control treatment, the fruit was washed under running water and then immersed in a chlorine solution of 150 mg L^{-1} for 10 minutes. After being dried at room temperature, all the fruit was stored in a cold room under conditions of 15 ± 0.2°C and 90 ± 2% RH. In all comparisons, the samples were analyzed every 3 days until 18 days of storage. In all treatments, the following analyses were carried out: peel and pulp chromatology, loss of fresh mass, soluble solids (SS), titratable acidity (TA), pH, ascorbic acid (AA), reducing sugars, respiration rate, specific polyphenoloxidase (PPO) and peroxidase (POD) activity, total antioxidant activity and total phenolic compounds. The experiment used a completely randomized design (DIC) in a 7x7 factorial scheme (organic acid x storage period). The data was subjected to analysis of variance using the F test at a 95% probability level, the means were compared using the Tukey test and regression was performed for the analysis of storage time. The results showed that the treatments with citric acid at 1% and 3% showed a climatic peak in the atemoya on the 6th day with (11.412 and 14.380 mL of co_2 Kg of $fruit^{-1}$ $hour^{-1}$), respectively 52.10% and 39.65% lower than the control treatment; the efficiency of organic acids in controlling bioactives that act on enzymatic browning cannot be proven.

Keywords: *Annona x atemoya* Mabb., ascorbic acid, citric acid, browning.

1.1 INTRODUCTION

The atemoia is a climacteric fruit Mosca (2006), which, due to its physiological characteristics, maintains its biochemical ripening process even after harvesting, a factor which influences its storage period. This statement corroborates Chitarra and Chitarra (2005), who explain that post-harvest losses in fruit occur due to damage resulting from problems during harvesting, transportation, storage and/or processing.

In processing technology, organic acids are used because they play an important role in controlling enzymatic browning, information that corroborates Wong (1994); Lima et al. (2011), who cite the use of organic acids (ascorbic acid and citric acid) as alternatives aimed at avoiding or minimizing the browning of tissues, increasing their useful life, as well as the loss of aroma, flavour and texture, a statement also confirmed by Souza, Correia; Santos (2010) who cite the organic acids present in foods as influencing the flavour, odour, color, stability and maintenance of food quality.

Excessive free radicals can generate oxidative stress in the metabolism of composite fruits (fruits that originate from the receptacle of a single flower, but with many ovaries); and the action of antioxidants, which are substances capable of inhibiting or preventing the oxidation of food Oetterer (2010), which protect cells against the damaging effects of oxygen and nitrogen free radicals formed in oxidative processes, which cause enzymatic browning in food (ALI et al., 2008; SOARES et al., 2005; GAVA, 2009).

Darkening can be controlled by chemical methods using antioxidants, which can be synthetic or natural, such as ascorbic acid, citric acid, calcium chloride and ethylene diamine tetraacetic acid; Examples of citric and ascorbic acids, either in an integrated or pure form, in the presence of quinones, which when oxidized, turn into meladoinins (browning) and in the presence of these acids return to their old form of phenolic compounds (FAGUNDES; AYUB, 2005; DUARTE ALMEIDA et al.,2006; EVANGELISTA, 2008).

The antioxidant action can theoretically take place by various means: deactivating active forms of oxygen, donating a hydrogen atom to the free radical and thus ending the oxidative process or slowing down the auto-oxidative process until all the antioxidant has been used, or donating a hydrogen to a peroxide free radical to form a hydroperoxide and an antioxidant free radical (OETTERER, 2010).

Regarding the mechanism of action of organic acids, Russel (1992) explains that the undissociated form of the acid is able to cross the cell membrane. Inside the cell, the acid would dissociate, altering the cytoplasmic pH, changing the proton gradient and the electrical charge with the outside of the cell. This would interfere with the amino acid and phosphate transport system and inactivate enzymes.

The growing interest in natural antioxidants from plant extracts is due to their low toxicity compared to synthetic antioxidants. Extracts from fruit, vegetables, cereals and their industrial by-products are rich in antioxidants, such as ascorbic acid, tocopherols, carotenoids and phenolic compounds (WOLFE; WU; LIU, 2003).

Cut fruit surfaces can be protected from browning by immersion in a 1-2% solution of citric acid. However, it is not always possible to lower the pH of the food due to the resulting acidic taste, which is incompatible with the sensory and technological properties of the final product (ARAUJO, 2012).

In a study on the most efficient methods for inhibiting enzymatic browning in fruit and vegetables, it was concluded that reducing the pH with 1.0% ascorbic acid was effective in inhibiting enzymatic browning (SOUZA et al., 2012).

Citric acid is an example of a natural organic acid (a chelating agent) which is used synergistically with ascorbic or erythorbic acids to prevent enzymatic browning caused by the action of the enzymes polyphenol oxidase and peroxidase (CHITARRA; CHITARRA, 2005; WILEY, 1994).

In a study on the preservation of minimally processed apples, they were cut and immersed in a preservative solution of 1% ascorbic acid and 0.5% citric acid for 2 minutes, then kept refrigerated at 2°C for 13 days; these treatments proved to be efficient in whitening and stiffening the pulp, without de-characterizing its natural condition (FONTES, 2005).

The rate of degradation of ascorbic acid is independent of pH, except in the range between 3 and 4, where there is a slight increase in oxidation. Accelerators of this reaction include fructose, fructose-6-phosphate, fructose-1,6-diphosphate, sucrose and caramelized fructose, with the final product of the reaction being furfural and CO_2 (ARAUJO, 2012).

Vitamin C is the most easily degradable of all the vitamins. It is only stable in acidic media and in the absence of light, oxygen and heat. The main factors capable of degrading ascorbic acid are: an alkaline environment, oxygen, heat, the action of light, metals (Fe, Cu, Zn) and the ascorbic acid oxidase enzyme (OLIVEIRA et al., 1999).

The aim of this experiment was to evaluate the behavior of ascorbic and citric acids in maintaining the quality of atemoya fruits in the post-harvest phase, when they were stored for 18 days under conditions of 15 ± 0.2°C and 90 ± 2% RH.

1.2 MATERIAL AND METHOD

The 'Thompson' cv. atemoias were purchased from a rural property located in Campos de Holambra, a district of the city of Paranapanema SP; the orchard is five years old and geographically located at the coordinates: latitude 22° 37' 59" south and longitude 47° 03' 20" west; at an altitude of 670 meters.

The atemoias were harvested in May, at dusk on the day before the analysis began. They were transported by pick-up truck at dawn on the day after the harvest, with the fruit previously selected and classified in terms of health and size. They were then stored in plastic boxes measuring 36 cm wide x 55 cm long and 30 cm deep, protected in layers by Kraft paper. The reference size and weight of the atemoias was standard 12 (twelve fruits per standard sales box). The standard box holds 4 to 4.5 kg of fruit, which determines the numerical reference cited. The physical, biochemical and enzymatic analyses were carried out in the post-harvest laboratory of the Faculty of Agronomic Sciences - FCA - "Júlio de Mesquita Filho" UNESP - Botucatu Campus, São Paulo; all the atemoias were immersed in chlorine solution at 150 mg L^{-1} , for 10 minutes and then dried at room temperature, before being treated with the aforementioned organic acids and their concentrations.

There were six treatments with organic acids (ascorbic and citric), plus the control treatment: TO - Control, T1 - 1% ascorbic acid, T2 - 2% ascorbic acid, T3 - 3% ascorbic acid, T4 - 1% citric acid, T5 - 2% citric acid and T6 - 3% citric acid.

Each treatment was assigned three replicates with two fruits per replicate, which were immersed in their respective organic acid solutions (ascorbic or citric) for 10 minutes and then dried at room temperature (Photo 1).

The atemoias were then packed in expanded polystyrene trays (2 fruits/tray) and sealed with 0.020mm polyvinyl chloride film (Photo 2) and stored in a cold room at 15 ± 0.2°C and 90 ± 2% RH for 18 days.

During the storage period, laboratory analyses were carried out every three days (0, 3, 6, 9, 12, 15 and 18).

Photo 1 - Atemoya cultivar 'Thompson' treated with organic acids for 10 minutes and then dried at room temperature.

Photo 2 - Atemoya cultivar 'Thompson' treated with organic acids and packed in expanded polystyrene trays sealed with 0.020mm polyvinyl chloride film.

The methodologies used for the physical, biochemical and enzymatic analyses are described below;

1.2.1 Mass Loss

Mass loss was measured using a MARTE® semi-analytical balance, model BL 3200 H - maximum load 2000g and precision 0.01g. The treatments were weighed at the beginning of the experiment and every 3 days for a total of 18 days, allowing the loss of fresh mass to be calculated as a percentage.

1.2.2 Hydrogen potential (pH)

The hydrogenic potential (pH) was measured from the crushed fruit pulp using a peagameter (Digital DMPH-2), according to the AOAC technique (1992),

1.2.3 Titratable acidity (TA)

The titratable acidity was expressed in grams of citric acid per 100g of pulp, obtained by titrating 5g of homogenized pulp diluted in 100 mL of distilled water with a standardized solution of sodium hydroxide at 0.1 Mol L^{-1} , using phenolphthalein as an indicator, which occurs when the parameter reaches 8.1, as recommended by the Adolfo Lutz Institute (BRASIL, 2008).

1.2.4 Soluble solids (SS)

Soluble solids were determined by refractometry in a digital refractometer Palette PR - 32, brand ATAGO, with automatic temperature compensation, according to AOAC (1992) and the results were expressed in °Brix.

1.2.5 Ascorbic Acid

Ascorbic acid was expressed as (mg ascorbic acid 100 g^{-1} pulp), values determined according to the method of Carvalho et al. (1990), which is based on the reduction of the indicator 2,6- dichlorobenzeneindophenol (DCFI) by ascorbic acid.

1.2.6 Reducing sugars (RA); total sugars (TA); starch and sucrose

For the quantification of reducing sugars (RA); total sugars (TA); starch and sucrose, the methodology used was described by Nelson (1944), adapted by Somogy (1945). The apparatus used was a Micronal B 382 spectrophotometer, and the readings were taken at 535 ηm, with the results expressed as a percentage.

1.2.7 Respiration rate

The respiration rate was determined indirectly in a respirometer by measuring the CO_2 released, according to the methodology adapted from Bleinroth et al. (1976). The respiration rate is calculated using the formula (1).

$$\frac{T\ CO2 = 22.(B-T).V1}{P.T.V2} \qquad (1)$$

Where:

T CO_2 = Respiration rate in grams of CO_2 kg^{-1} of fruit hour ;$^{-1}$

B = Volume of standardized HCl (mL) used to titrate the potassium hydroxide solution before contact with the CO_2 produced by the fruit;

T = Volume of standardized HCl (mL) used to titrate the potassium hydroxide after contact with the CO_2 produced by the respiration of the fruit;

V1 = Volume of 0.1 M KOH (mL) used for CO_2 absorption (mL);

V2 = Volume of 0.1 KOH (mL) used for titration (mL) with HCL;

P = Fresh mass (kg);

T = Reaction time (hour);

22 = Inherent in the CO_2 equivalent = 44 / 2, multiplied by the concentration of hydrochloric acid

1.2.8 Instrumental color evaluation

Coloration was determined by measuring two external points on the peel, one on each side of the fruit, and two internal points on the flesh, using the same methodology; the action was carried out using a Konica Minolta colorimeter (Chroma meter, CR 400), determining the values (L*, *C** and *H°*). Where L*, expressed as a percentage, indicates luminosity values (0% = black and 100% = white). C* is represented by Chroma, which defines the intensity of the color.

°*Hue* is the value in degrees corresponding to the three-dimensional color diagram 0° (red), 90° (yellow) and 270° (blue). °*Hue* ranges from 0 to 18° for red-violet, 19 to 54° for red, 55 to 90° for orange, 91 to 126° for yellow, 127 to 162° for yellow-green, 163 to 198° for green, 199 to 234° for blue-green, 235 to 270° for blue, 271 to 306° for blue-violet and 307 to 342° for violet, 343 to 360° red-violet, making 360° (MINOLTA, 1998).

1.2.9 Specific activity of Polyphenoloxidase - PPO (EC. 1.14.18.1)

The extraction and determination of the enzyme polyphenoloxidase (PPO) - (EC. 1.14.18.1), was carried out according to the method of Cano et al. (1997) and adaptations. The results are expressed as UAE^{-1} min^{-1} g^{-1} of fresh matter. To prepare the enzyme extract, 0.5 g of atemoya was weighed, crushed and mixed with 10 mL of 2M potassium phosphate buffer pH 6.0. The sample was centrifuged for 15 minutes at 6000 rpm at 4°C. The extract was filtered and the samples pipetted with 1.85 mL of 0.1 M catechol and 0.3 mL of the extract. To zero the spectrophotometer, 0.3 mL of 100 mM sodium acetate buffer and 1.85 mL of 0.1 M catechol were pipetted in. After this process, it was placed in a water bath for 30 minutes at 30°C. Then 0.8 mL of perchloric acid was added to the tubes containing the samples to stop the reaction and the spectrophotometer was read at a wavelength of 460 ηm.

1.2.10 Specific activity of Peroxidase - POD (EC. 1.11.1.7)

The extraction and determination of the enzyme peroxidase (POD) - (EC. 1.11.1.7), was done according to adaptations to the method of Lima et al. (1999). The results are expressed as UAE^{-1} min^{-1} g^{-1} of fresh matter. To prepare the enzyme extract, 0.5 g of atemoya was weighed, crushed and mixed with 10 mL of 2M potassium phosphate buffer pH 6.0. The sample was centrifuged for 15 minutes at 6000 rpm at 4°C. The extract was filtered and pipetted with 0.5 mL of peroxide, 0.5 mL of aminoantiperine, 1 mL of extract, placed in a water bath at 30°C for 5 minutes and 2 mL of PA ethyl alcohol was used to stop the reaction. When crystals formed, the water bath was boiled for 5 minutes. To zero the spectrophotometer, 0.5 mL of peroxide, 0.5 mL of Aminoantiperine, 1 mL of potassium phosphate cap and 2 mL of ethyl alcohol were pipetted in. Reading was carried out on a spectrophotometer at a wavelength of 460 ηm.

1.2.11 Antioxidant activity and total phenolic compounds

For the evaluation of antioxidant activity and total phenolic compounds, the methanolic extract of the pulp was prepared using the solvent mixture methanol: water (70:30 v/v) for the extraction, as this was the most suitable solvent for the extraction of phenolic compounds among those tested (alcohol, water and acetone). The extracts were obtained in triplicate. 0.2 g of the fruit pulp was weighed into Falcon tubes and 10 mL of the methanol: water mixture (70:30 v/v) was added. The tubes containing the sample and the solvent were submitted to Turrax for a few minutes at room temperature. The extracts were then placed in an ultrasonic bath for 20 minutes and then centrifuged at 6000 rpm for 15 minutes. The extracts were then stored in amber bottles at a temperature of 8°C until the time of analysis, and for no longer than seven days (MENSOR et al., 2001; (SINGLETON; ORTHOFER; LAMUELA, 1999).

1.2.11.1 Total antioxidant activity

In the evaluation of total antioxidant activity, the sequestering capacity was determined using the DPPH (1,1-diphenyl-2-picrylhydrazyl) method, based on the principle that DPPH is a stable violet radical which accepts an electron or a hydrogen radical to become a stable molecule, being reduced in the presence of an antioxidant and acquiring a yellow color. In radical form, DPPH has a characteristic absorption at 517 ηm, which disappears as it is reduced by the hydrogen donated by an antioxidant compound (MENSOR et al., 2001). The reaction mixture was made up of 500µL of the methanolic extracts of the pulp, 3.0mL of 99% methanol and 300µL of DPPH radical in 0.5mM methanol solution and incubated for 45 minutes at room temperature and protected from light. Expressed as a percentage.

1.2.11.2 Total phenolic compounds

The total content of phenolic compounds in the methanolic extract of the pulp was determined using the Folin-Ciocalteau spectrophotometric method (SINGLETON; ORTHOFER; LAMUELA, 1999). The Folin-Ciocalteau reagent is a complex solution of polymeric ions formed from phosphomolybdic and phosphotungstic heteropoly acids. This reagent oxidizes the phenolates, reducing the acids to a blue Mo-W complex. It was read in a spectrophotometer at 760 ηm. To carry out the analysis, a 0.2 mL aliquot of the methanolic extract was transferred to a tube and 2.5 mL of Folin Ciocalteau reagent was added, diluted 1:10 in water. The mixture was left to stand for 5 minutes. Then 2mL of 4% sodium carbonate was added and the tubes were left to stand for 2 hours, protected from light. The absorbance was measured in a UV-BEL PHOTONICS SP-1105 spectrophotometer at 760ηm. A blank sample was taken under the same conditions and the results of the total phenolic compounds were expressed in mg of gallic acid 100g-1, based on a calibration curve of gallic acid with concentrations ranging from 5 to 100µg/mL.

1.2.12 Experimental Design and Statistical Analysis

The experimental design and statistical analysis followed the criteria of an inter-randomized design (ICD) in a 7 x 7 factorial scheme (organic acid x storage period) with three replications. The data was subjected to analysis of variance using the F test at a 95% probability level. The means were compared using Tukey's test and regression was performed for the analysis of storage time; the statistical program used was Sisvar (FERREIRA, 2010).

1.3 RESULTS/DISCUSSION

The respiratory activity of atemoya fruits treated with organic acids is shown in Figure 1. 'Thompson' atemoia fruits show climacteric behavior, as observed by Lundgren (2017) in fruits of the same cultivar. On the sixth day of storage, a climacteric peak was observed in all the treatments, and the treatments with citric acid at 1% and 3% were the ones with the lowest respiration rates respectively (11.412 and 14.380 mL of CO_2 Kg of fruit^{-1} hour^{-1}), representing 52.10% and 39.65% lower than the control treatment. This data corroborates Mosca (2003), who observed a respiration peak on the 7th day of storage for 'Gefner' atemoia cv. at a rate of 25.68 mg $C\theta_2.kg^{\wedge 1}.h^{\wedge 1}$, with a subsequent decrease on the 9th day of storage to a respiration rate of 20.65 mg $C\theta_2.kg^{\wedge 1}.h^{\wedge 1}$; this difference may be associated with the difference in treatment between the experiments; while Mosca (2003) stored the atemoyas in containers with atmospheric air circulation, this experiment treated the atemoyas with organic acids and stored them at a temperature of 15 ± 0.2°C and 90 ± 2% relative humidity (RH). The characterization of respiration can be explained by studies carried out by Bruinsma (1984) and Anese; Fronza (2015) who suggest that in the post-harvest physiology of fruit, the climacteric peak is characterized by transforming actions that include reducing acidity, reducing starch content, reducing astringency, increasing total sugars, increasing aromas and degrading chlorophyll; This information makes it possible to interpret that the lower respiratory activity allows the fruit to delay its ripening process.

Figure 1 - Respiratory activity (mL of CO_2 Kg of fruit^{-1} hour^{-1}) in 'Thompson' cv. atemoia, treated with organic acids and packed in expanded polystyrene and polyvinyl chloride trays and stored at 15 ± 0.2°C and 90 ± 2% RH for 18 days. Botucatu SP

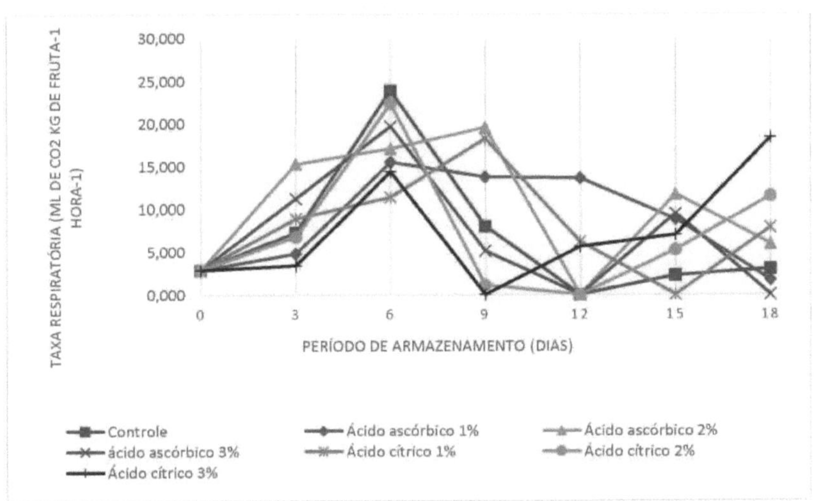

In the data analysis, which characterizes the loss of mass in (%), in the atemoias cv. 'Thompson' treated with organic acid, and evaluated during 18 days of storage, can be seen in Table 1; which shows that there was a statistical difference between the control treatment and the treatment with 1% ascorbic acid, which did not differ statistically from the treatments with 2 and 3% ascorbic acid and the treatment with 2% citric acid; these data differ from those found by Silva (2011), who evaluated mass loss in atemoya cv. 'Gefner', stored at a temperature of 25°C and analyzed over periods of 0, 3, 6 and 9 days, showed losses of 12.38%; and also diverges from Lundgren (2017), who observed values of 8.5 to 12% loss of fresh mass in atemoia cv. 'Thompson' treated with 1-MCP and stored for 18 days in a cold room at 15°C ± 1 and 90 ± 5% RH; the respective differences may be related to the storage temperature (25°C) in the first case and the treatment differential adopted in the second case (1-MCP).

Table 1 - Mass loss (%) in 'Thompson' cv. atemoia treated with organic acids and packed in expanded polystyrene and polyvinyl chloride trays and stored at 15 ± 0.2°C 90 ± 2% RH for 18 days. Botucatu SP

Treatment	Storage period (days)							Average
	0	3	6	9	12	15	18	
Control*	0,00	0,24	0,41	0,61	0,83	1,14	1,40	0.66 bc
Ascorbic 1%	0,00	0,24	0,49	0,78	1,15	1,66	2,01	0,91 a
Ascorbic 2%	0,00	0,28	0,46	0,75	1,09	1,51	1,95	0.86 ab
Ascorbic 3%	0,00	0,27	0,46	0,65	0,90	1,17	1,41	0.69 abc
Citrus 1%	0,00	0,17	0,34	0,54	0,77	1,05	1,28	0,59 c
Citric 2%	0,00	0,31	0,53	0,73	0,97	1,24	1,41	0.74 abc
Citrus 3%	0,00	0,20	0,36	0,57	0,80	1,08	1,37	0,62 c
C.V. (%)	32,0							

Averages followed by the same lower-case letter in the column do not differ significantly from each other at 5% probability, according to Tukey's test.

(*) The control treatment is characterized by the fruit being washed in running water and then immersed in a chlorine solution at 150 mg L^{-1} for 10 minutes and then dried at room temperature.

Figure 2 shows an upward trend in the averages for mass loss (%) for all treatments during the 18 days of analysis; this is related to the loss of water and solutes during ripening, reaching a maximum percentage loss of 1.54% compared to day zero; This result makes it possible to interpret that the levels of loss obtained meet the requirements of Chitarra and Chitarra (2005), who state that losses of between 3 and 6% are likely to cause a sharp decline in the quality of plant products, which was not the case in this study.

Figure 2 - Mass loss (%) in 'Thompson' cv atemoia, treated with organic acids and packed in expanded polystyrene and polyvinyl chloride trays and stored at 15 ± 0.2°C and 90 ± 2% RH for 18 days. Botucatu SP

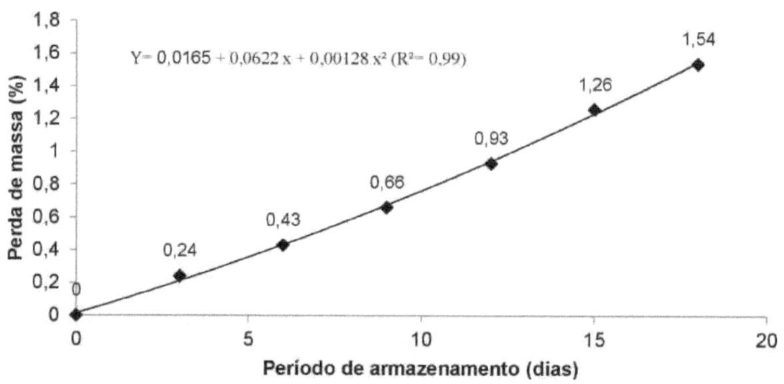

The analysis of soluble solids shows that during the 18 days of storage there was no statistically significant variation for the treatments or for their interaction with storage, which makes it possible to interpret that the use of organic acids did not interfere with the variation in soluble solids of the atemoya during its natural ripening process (Table 2).

Table 2 - Soluble solids (°Brix) in atemoia cv. 'Thompson', treated with organic acids and packed in expanded polystyrene and polyvinyl chloride trays and stored at 15 ± 0.2°C and 90 ± 2% RH for 18 days. Botucatu SP

Treatment -	Storage period (days)							Average
	0	3	6	9	12	15	18	
Control *	13,83	17,10	21,37	20,17	20,60	22,70	24,00	19,97
Ascorbic 1%	13,83	17,17	19,97	21,77	19,63	23,20	23,83	19,91
Ascorbic 2%	13,83	19,47	21,37	23,20	22,87	19,63	23,80	20,59
Ascorbic 3%	13,83	14,40	19,03	25,53	20,70	22,33	25,77	20,23
Citrus 1%	13,83	19,03	20,37	21,77	21,70	20,33	23,20	20,03
Citric 2%	13,83	15,73	21,07	22,90	23,67	21,70	24,97	20,55
Citrus 3%	13,83	18,77	25,90	22,73	23,77	24,43	22,70	21,73
C.V. (%)	11,7							

The means of treatment and interaction were not significant at 5% probability, according to Tukey's test.

(*) The control treatment is characterized by the fruit being washed in running water and then immersed in a chlorine solution of 150 mg L^{-1} for 10 minutes and then dried at room temperature.

The variation in the percentages of soluble solids obtained in the experiment can be seen in Figure 3; the range of variation between day zero 13.8 ° Brix and day 18 18.24 ° Brix was 32%; this variation may characterize an important variable for the industrial processing of the atemoya; this statement corroborates Pinheiro et al, (1984), who characterize high levels of soluble solids in fruit as an economic factor in the addition of sugars, less time spent evaporating water, lower energy expenditure and greater product yield during processing; and according to Maia et al, (1986), the soluble solids content is also a parameter that has been used as an indicator of the quality of pine cone fruit, the data obtained in this experiment differs from that found by Lundgren (2017), who analyzed 'Thompson' cv. atemoia and found soluble solids values ranging from 28.6 °Brix to 32 °Brix at the end of 18 days of storage, which corroborates Silva (2011), who found soluble solids values ranging from 10 to 32.15^0 Brix, for 'Gefner' cv. atemoia. 'Gefner'; while Neves (2009) found soluble solids values of 26.1° Brix in atemoia cv. 'Thompson' grown in the northern region of the state of Paranà, which is close to the maximum value found in this experiment and corroborates the soluble solids value found by Souza (2013) in atemoia cv. 'Gefner' of 25.95 °Brix; these data are also close to those found by Mosca; Cavalcante; Datas (2006), who characterized the fruit of the atemoia, in Brazilian conditions, with soluble solids contents ranging from 4 °Brix to 27 °Brix.

Figure 3 - Soluble solids (° Brix) in 'Thompson' cv. atemoia, treated with organic acids and packed in expanded polystyrene and polyvinyl chloride trays and stored at 15 ± 0.2°C and 90 ± 2% RH for 18 days. Botucatu SP

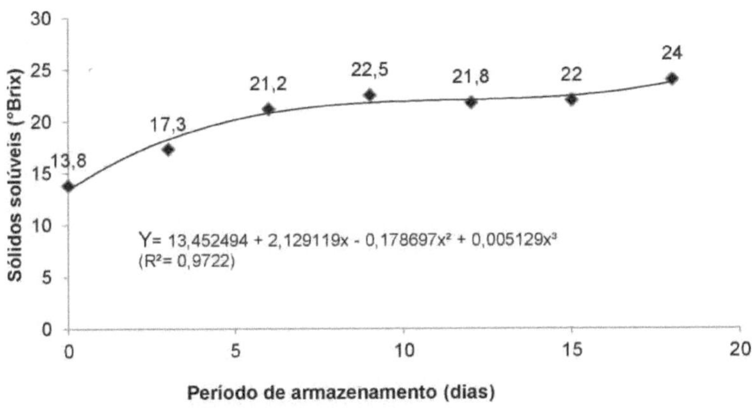

Table 3 shows the variation in pH values of the fruit of the 'Thompson' atemoia cv. The control treatment and the treatment with 1% ascorbic acid differed statistically from the treatment with 3% citric acid, showing pH values of 5.7, 5.71 and 5.47 respectively; The other treatments did not differ statistically from the 3% citric acid treatment. This result corroborates Braverman (1967), who mentions the importance of lowering pH values during the fruit ripening process, in order to reduce the speed at which the fruit darkens, thus improving its visual appearance in the post-harvest period; This statement corroborates Anese; Fronza (2015), who characterize the presence of organic acids in fruit as an influencing factor in ripening, these being direct influencers of the pH value of the medium; these statements characterize the importance of the study related to the analysis of the variation in pH of the fruit during storage, as a way of interpreting the importance of a technology in increasing the shelf life of the fruit.

Table 3 - pH values of 'Thompson' cv. atemoia, treated with organic acids, packed in expanded polystyrene and polyvinyl chloride trays and stored at 15 ± 0.2°C and 90 ± 2% RH for 18 days. Botucatu SP

Treatment	Storage period (days)							Average
	0	3	6	9	12	15	18	
Control *	5,76	5,86	5,49	5,68	5,69	5,82	5,65	5,7 a
Ascorbic 1%	5,76	5,99	5,94	5,54	5,66	5,63	5,50	5,71 a
Ascorbic 2%	5,76	5,68	5,59	5,48	5,69	5,47	5,61	5.61 ab
Ascorbic 3%	5,76	5,67	5,65	5,47	5,72	5,56	5,59	5.63 ab
Citrus 1%	5,76	5,63	5,68	5,48	5,66	5,43	5,45	5.58 ab
Citric 2%	5,76	5,52	5,73	5,58	5,54	5,46	5,46	5.58 ab
Citrus 3%	5,76	5,62	5,09	5,41	5,50	5,53	5,38	5,47 b
C.V. (%)	3,1							

Averages followed by the same lower-case letter in the column do not differ significantly from each other at 5% probability using the Tukey test. (*) The control treatment is characterized by the fruit being washed in running water and then immersed in a chlorine solution at 150 mg L^{-1} for 10 minutes and then dried at room temperature.

The data obtained in the experiment for the variation in pH during the 18 (eighteen) days of analysis can be seen in Figure 4, which shows a non-significant variation in this period (5.76 to 5.52); this data may characterize a longer shelf life for the atemoia in post-harvest; but the pH of this period is contrary to that found by Mosca (2002), who studied atemoia cv. 'Gefner' stored at 27°C, observed that between day zero and day 6 there was a variation in pH between 5.5 and 4.25; the explanation for the disagreement in values may be related to the storage temperature, which differs from this experiment.

Figure 4 - pH values in 'Thompson' cv. atemoia, treated with organic acids and packed in expanded polystyrene and polyvinyl chloride trays and stored at 15 ± 0.2°C and 90 ± 2% RH for 18 days. Botucatu SP

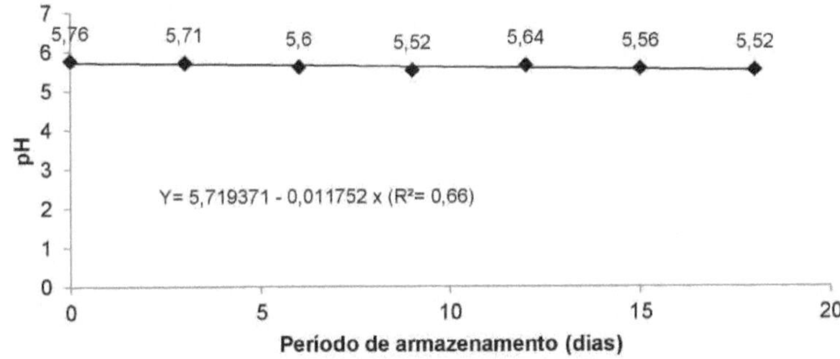

Table 4 shows the titratable acidity data for the different treatments and their interaction with the storage period. The results show that there was no statistical significance between the treatments and their interaction with storage, which makes it possible to interpret that the use of organic acids did not interfere with the titratable acidity differential during the natural ripening process of the atemoia.

Table 4 - Titratable acidity (g citric acid 100 g^{-1}) in 'Thompson' cv. atemoia, treated with organic acids and packed in expanded polystyrene and polyvinyl chloride trays and stored at 15 ± 0.2°C and 90 ± 2% RH for 18 days. Botucatu SP

Treatment	Storage period (days)							Average
	0	3	6	9	12	15	18	
Control *	0,136	0,123	0,143	0,127	0,120	0,150	0,107	0,129
Ascorbic 1%	0,136	0,127	0,117	0,117	0,130	0,150	0,137	0,130
Ascorbic 2%	0,136	0,130	0,133	0,157	0,163	0,180	0,110	0,144
Ascorbic 3%	0,136	0,120	0,127	0,163	0,143	0,150	0,133	0,139
Citrus 1%	0,136	0,123	0,107	0,107	0,147	0,173	0,163	0,137
Citric 2%	0,136	0,137	0,110	0,107	0,150	0,123	0,137	0,128
Citrus 3%	0,136	0,117	0,210	0,127	0,170	0,153	0,130	0,149
C.V. (%)	23,9							

The means of treatment and interaction were not significant at 5% probability, according to Tukey's test.

(*) The control treatment is characterized by the fruit being washed under running water and then immersed in a chlorine solution of 150 mg L^{-1} for 10 minutes and then dried at room temperature.

The ascorbic acid content observed in Figure 5 and measured in mg of ascorbic acid 100 g^{-1} , showed a linear and constant drop, with the exception of the treatments with 3% citric acid, which showed the highest index on the 3rd day of storage with 30,5 mg of ascorbic acid 100 g^{-1} and the treatment with 1% ascorbic acid, which showed the lowest index of 4.5 mg of ascorbic acid 100 g^{-1} , for the same period. This relationship model is repeated on the 6th and 9th day of analysis; this model is inverted on the 12th and 15th day of storage, when the treatment with 1% ascorbic acid shows higher values than the treatment with 3% citric acid; the importance of the higher concentration of ascorbic acid during the storage period corroborates Levine (2011), who characterizes the importance of vitamin C as an enzyme cofactor; inactivating free radicals, making it an excellent antioxidant; and thus favoring a greater durability of the food during its storage, the results obtained in this experiment these data differ from Lundgren (2017) who obtained 0.45% titratable acidity in atemoya cv. 'Thompsom', treated with 1-MCP on the 9th day of analysis; this difference may be associated with the use of 1-MCP, as this experiment adopted treatment with organic acids; data obtained by Silva (2011), in atemoya cv. 'Gefner' atemoyas kept at room temperature (24.8 ± 1.0°C and 48 ± 11% RH), also differed from this experiment, as their averages were 0.40% for 'fresh' fruit and 0.33% for 'ripe' fruit; this difference may be associated with the storage conditions of the fruit; as in this experiment the storage conditions were 15 ± 0.2°C and 90 ± 2% RH; these high average data are characterized by Rocha et al, (2001), that from an industrial point of view, natural acidity in food reduces the need to add acidifiers and provides nutritional improvement, food safety and organoleptic quality. Among the treatments, the highest average found was with citric acid at 3% (0.149%), this average is 49.6% lower than that found by Neves (2009), in 'Thompson' cv. atemoia, planted at 480 meters above sea level, in the northern region of the State of Paranà; this altitude may have influenced the result, since in this experiment the atemoia was planted at 670 meters above sea level, in the region of Paranapanema S.P.

Figure 5 - Ascorbic acid content (mg ascorbic acid $100g^{-1}$) in atemoya cv. 'Thompson', treated with organic acids and packed in expanded polystyrene and polyvinyl chloride trays and stored at 15 ± 0.2°C and 90 ± 2% RH for 18 days. Botucatu SP

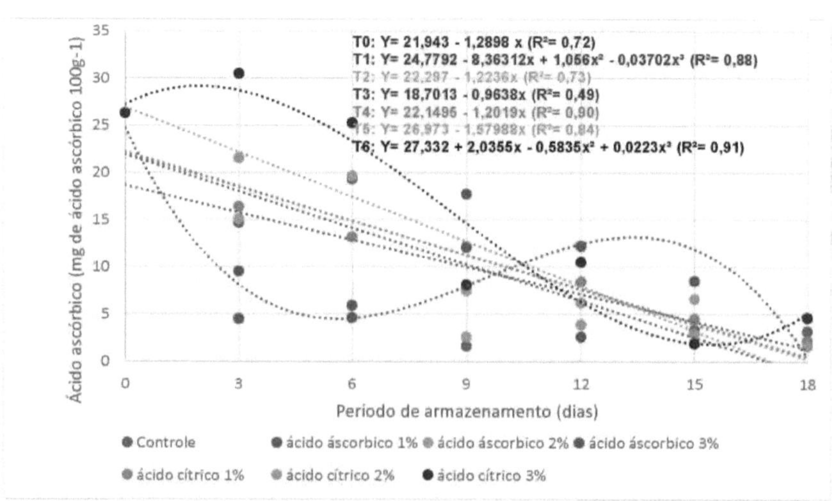

The analysis of the data obtained for reducing sugar in Table 5 showed that there was no statistically significant difference between the treatments and their interaction with the

storage period, which makes it possible to interpret that the organic acids did not interfere with the natural ripening process of the atemoia in terms of this item.

Table 5 - Reducing sugar (%) in 'Thompson' cv. atemoia, treated with organic acids and packed in expanded polystyrene and polyvinyl chloride trays and stored at 15 ± 0.2°C and 90 ± 2% RH for 18 days. Botucatu SP

Treatment	Storage period (days)							Average
	0	3	6	9	12	15	18	
Control *	3,03	3,91	4,96	5,99	6,35	9,50	9,78	6,22
Ascorbic 1%	3,03	3,29	4,69	5,83	6,26	9,33	10,74	6,17
Ascorbic 2%	3,03	3,60	5,92	6,81	4,80	7,12	14,16	6,49
Ascorbic 3%	3,03	4,19	5,75	5,87	6,79	9,09	8,48	6,17
Citrus 1%	3,03	3,58	5,08	6,38	5,68	8,36	11,53	6,23
Citric 2%	3,03	4,22	5,43	5,84	10,70	8,33	8,74	6,61
Citrus 3%	3,03	4,29	7,12	6,81	9,02	9,54	10,57	7,20
C.V. (%)	30,1							

The means of treatment and interaction were not significant at 5% probability, according to Tukey's test.

(*) The control treatment is characterized by the fruit being washed in running water and then immersed in a chlorine solution of 150 mg L^{-1} for 10 minutes and then dried at room temperature.

The data obtained with reducing sugar referring to the averages of the storage periods are shown in Figure 6, which makes it possible to observe that there was a minimum of 3.03% at the beginning of the experiment on day zero and a maximum value of 14.16% on the 18th day, a value reached by the treatment with 2% citric acid, characterizing 467% of variation over the 18-day storage period; the index obtained on the 18th day of storage is close to that obtained by Lundgren (2017), who obtained 15.38% in 'Thompson' taro. 'Thompson', treated with different doses of 1- MCP; it is also close to that found by Alves et al. (2001), who analyzed pine cones at room temperature until they were fully ripe and obtained 15.96% reducing sugar; Nogueira et al. (2005) characterizes the increase in the concentration of reducing sugar as a consequence of the hydrolysis of starch and the inversion of sucrose by glucose and fructose, which are indicative of ripening in the fruit.

Figure 6 - Reducing sugar content (%) in 'Thompson' cv. atemoia, treated with organic acids, conditioned in expanded polystyrene and polyvinyl chloride trays and stored at 15 ± 0.2°C and 90 ± 2% RH for 18 days. Botucatu SP

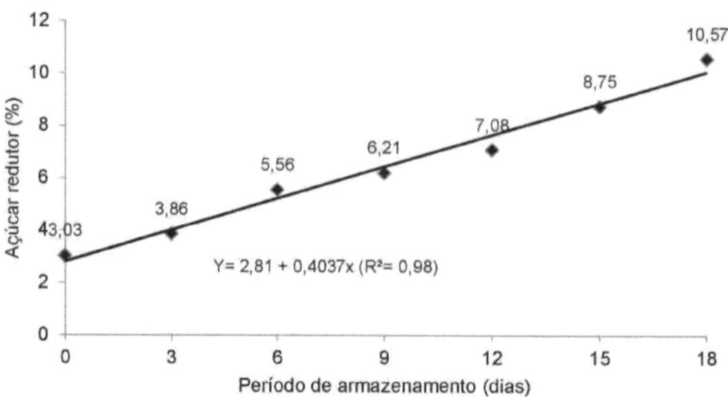

The total sugar index found in 'Thompson' cv. atemoia shows that there was no statistical significance in the treatments or in the interaction between treatments and storage period, which makes it possible to interpret that the organic acids did not interfere in the natural ripening process of the atemoia, as far as the total sugar criterion is concerned.

Table 6 - Total sugar (%) in 'Thompson' cv. atemoia, treated with organic acids and packed in expanded polystyrene and polyvinyl chloride trays and stored at 15 ± 0.2°C and 90 ± 2% RH for 18 days. Botucatu SP

Treatment	Storage period (days)							Average
	0	3	6	9	12	15	18	
Control *	5,15	7,56	15,55	8,70	11,34	12,90	15,65	10,98
Ascorbic 1%	5,15	7,54	11,95	9,26	10,54	15,95	16,94	11,05
Ascorbic 2%	5,15	7,81	12,79	10,48	6,98	11,98	17,94	10,45
Ascorbic 3%	5,15	7,68	10,39	9,05	10,36	15,79	16,08	10,64
Citrus 1%	5,15	8,41	10,31	8,23	11,50	14,79	16,16	10,65
Citric 2%	5,15	9,29	10,75	8,07	15,39	14,96	16,13	11,39
Citrus 3%	5,15	8,95	17,65	9,50	15,49	17,72	16,78	13,03
C.V. (%)	25,1							

The means of treatment and interaction were not significant at 5% probability, according to Tukey's test.

(*) The control treatment is characterized by the fruit being washed in running water and then immersed in a chlorine solution at 150 mg L^{-1} for 10 minutes and then dried at room temperature.

An analysis of the behavior of total sugars over the storage period shows an increase in total sugar from 5.174% on day zero to 16.52% on day 18; between day 6 and 9 there was a reduction in total sugars of around 28.8% (Figure 7). This reduction is related to a decrease in the respiratory action of 'Thompson' atemoya. 'Thompson' as ripening progresses, the concentration of soluble sugars such as glucose, fructose and sucrose increases, and there may be a subsequent decline due to their use as a source of energy (CHITARRA; CHITARRA, 2005). This phenomenon may be related to a decrease in the respiratory peak and ethylene (LIMA et al., 2003), hydrolysis of starch and changes in cellular appearance (GONI et al., 2010). These statements make it possible to interpret that there was a positive relationship between the organic acids used in this experiment and the behavior observed between the periods mentioned.

Figure 7 - Total sugar (%) in 'Thompson' cv. atemoia, treated with organic acids, packed in expanded polystyrene and polyvinyl chloride trays and stored at 15 ± 0.2°C and 90 ± 2% RH for 18 days. Botucatu SP

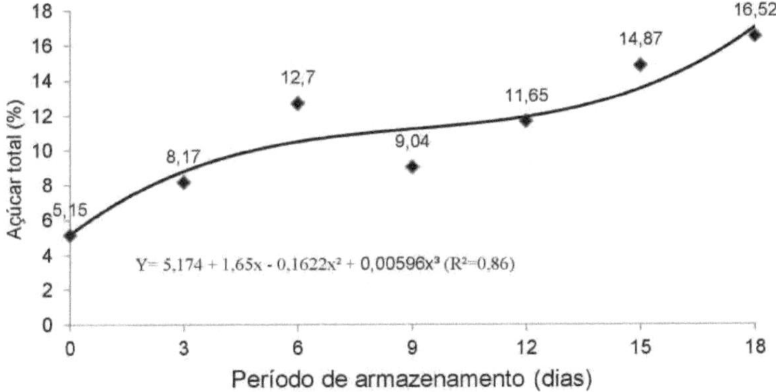

In the analysis of the results obtained in the experiment, regarding the concentration of sucrose in the 'Thompson' cv. atemoyas, we can see in Figure 8 that from the 6th to the 18th day of storage, the fruit from the treatment with 3% citric acid showed a trend, with higher sucrose indices than the other treatments; this data may favor a longer shelf life for the 'Thompson' cv. atemoyas. This information is based on the fact that sucrose can become a reducing sugar if it undergoes enzymatic action or acid hydrolysis, thus forming monosaccharides (glucose and fructose) during storage; Nogueira (2005) characterizes the variation in total sugar indices (glucose plus fructose), resulting from the conversion of sucrose, as an interference directly proportional to fruit ripening and Lima et al. (2010a), reports sucrose as the predominant sugar at the beginning of ripening, its formation resulting from the hydrolysis of starch, which precedes the accumulation of glucose and fructose, probably, the transformation of starch into sucrose, is the main pathway for the degradation of starch during ripening to this information Lundgren (2017), reports that treating taro cv. 'Thompson' with 1-MCP, obtained a sucrose peak on the 6th day of analysis and on the 12th day the value remained similar, data which is similar to that obtained in this experiment, in terms of the delay in sucrose hydrolysis, from the 6th to the 12th day of storage.

Figure 8 - Sucrose content (%) in 'Thompson' cv. atemoia, treated with organic acids and packed in expanded polystyrene and polyvinyl chloride trays and stored at 15 ± 0.2°C and 90 ± 2% RH for 18 days. Botucatu SP

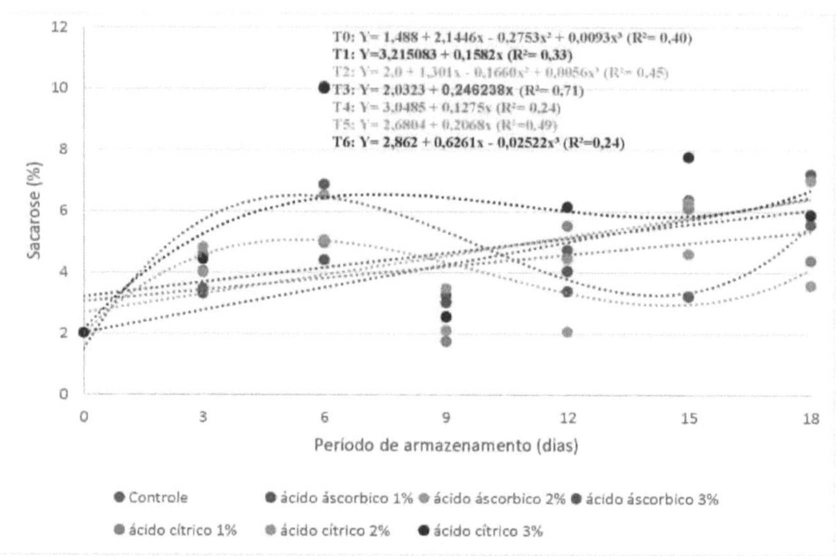

Table 7 shows that there was no interaction between the fruit, the treatments and the period of analysis with regard to the concentration of starch in 'Thompson' atemoias, making it possible to interpret that the use of organic acids had no influence on the changes in their content during the natural ripening process.

Table 7 - Starch values (%) in 'Thompson' cv. atemoia, treated with organic acids and packed in expanded polystyrene and polyvinyl chloride trays and stored at 15 ± 0.2°C and 90 ± 2% RH for 18 days Botucatu SP

Treatment	Storage period (days)							Average
	0	3	6	9	12	15	18	
Control *	18,69	11,87	9,78	12,18	10,52	8,45	5,61	11,01
Ascorbic 1%	18,69	14,31	12,05	14,69	11,57	5,68	2,17	11,31
Ascorbic 2%	18,69	14,68	10,61	10,60	13,71	9,00	3,24	11,50
Ascorbic 3%	18,69	14,82	12,46	10,40	7,14	5,28	3,00	10,25
Citrus 1%	18,69	14,51	14,85	13,55	10,23	5,99	4,24	11,72
Citric 2%	18,69	13,44	12,58	16,21	3,20	7,40	4,90	10,92
Citrus 3%	18,69	14,35	9,00	12,87	6,61	5,06	1,95	9,79
C.V. (%)	30,1							

The means of treatment and interaction were not significant at 5% probability, according to Tukey's test.

(*) The control treatment is characterized by the fruit being washed in running water and then immersed in chlorine solution at 150 mg L^{-1} for 10 minutes and then dried at room temperature.

The variation in starch indices shown in Figure 9 ranges from 18.69% on day zero to 1.95% on day 18; the latter value refers to ; this index is higher than that found by Alves et al. (2001) who, when evaluating pine cones at room temperature until they were fully ripe, obtained 0.87% starch; this difference may be related to the storage temperature of the pine cones at 25°C, which was higher than that used in this experiment 15 ± 0.2°C. The reduction in starch levels in the 3% citric acid treatment, as in the other treatments, can be explained, according to Aneze and Fronza (2015) and Chitarra and Chitarra (2005), by the conversion of starch (the main carbon reserve used to synthesize sucrose in the post-harvest period) into soluble sugars during fruit ripening; The starch content in the fruit is an indicator of its degree of ripeness; this statement explains Figure 9 with a reduction from 18.69% on day zero to 3.58% on the 18th day of analysis; These data corroborate Torres (2008), who characterizes anonaceae as fruits with a large reserve of starch in their fresh fruit mass (10 to 12%), the starch being hydrolysed after harvesting in the fruit; the relationship between starch and fruit ripening was also observed by Silva; Lajolo; Cordenunci (2003), in their analysis of the evolution of ripening in mango cultivars, compared the percentage variation in starch content, which showed a gradual and constant drop during the experiment, with a reduction of 19% in the interval. Figure 9 shows a steady and gradual drop in the starch content of 'Thompson' cv. atemoia fruit over the 18 days of storage; this drop can be explained by the hydrolysis of starch during the respiration process, transforming it into reducing sugars, and the behavior of the results can be related to the observations of Mota et al. (2002) and Nascimento et al, (2006), who state that various enzymes present in plant tissue are capable of metabolizing starch, especially α-amylase, β-amylase and starch phosphorylase. These enzymes show an increase in activity during fruit ripening, causing a slower transformation of starch into soluble sugars and thus delaying ripening.

Figure 9 - Starch content (%) in atemoya cv. 'Thompson', treated with organic acids and packed in expanded polystyrene and polyvinyl chloride trays and stored at 15 ± 0.2°C and 90 ± 2% RH for 18 days. Botucatu SP

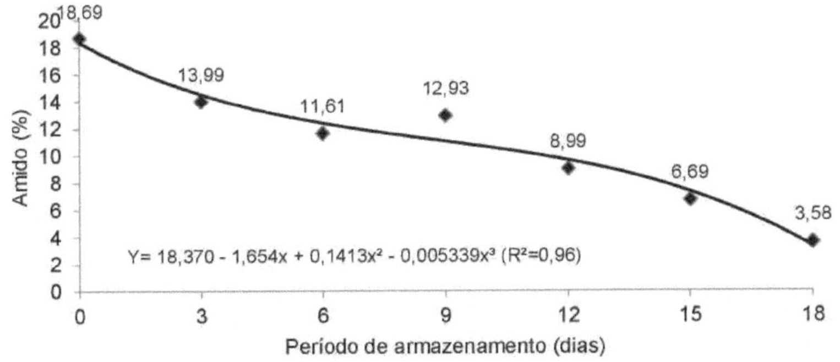

The values for the brightness of the peel of the atemoya fruit are shown in Table 8 below, which shows that there was no significance between the treatments and no significance in their interaction with storage time. This statement makes it possible to interpret the lack of influence of organic acids in maintaining the brightness of the peel during the natural ripening process. The importance of skin brightness can be a parameter used to characterize fruit ripeness, thus influencing the consumer's choice of products. This statement corroborates Teixeira (2013), who studied the characterization of factors for choosing and buying fruit and vegetables by the adult population of the Federal District and found that appearance influenced the choice and purchase of fruit and vegetables by 74.4%; This statement corroborates this experiment, characterizing the importance of evaluating the luminosity of fruit peels, with the aim of increasing their shelf life and consumer acceptance.

Table 8 - Luminosity values (L*) of the peel in 'Thompson' cv. atemoia, treated with organic acids and packed in expanded polystyrene and polyvinyl chloride trays and stored at 15 ± 0.2°C and 90 ± 2% RH for 18 days. Botucatu SP

Treatment	Storage period (days)							Average
	0	3	6	9	12	15	18	
Control *	56,6	52,2	55,2	50,8	53,5	53,6	51,0	53,3
Ascorbic 1%	56,6	55,2	52,5	51,7	53,3	54,5	54,4	54,0
Ascorbic 2%	56,6	53,3	50,3	53,1	53,3	52,3	42,3	51,9
Ascorbic 3%	56,6	54,8	53,7	51,9	53,5	54,1	56,1	54,4
Citrus 1%	56,6	52,9	53,5	52,2	47,9	55,1	52,4	52,9
Citric 2%	56,6	51,0	56,6	53,4	50,1	56,0	54,8	54,0
Citrus 3%	56,6	53,2	55,5	56,3	52,8	51,1	47,1	53,2
C.V. (%)	8,0							

The means of the treatment and interaction were not significant at 5% probability, according to Tukey's test.

(*) The control treatment consisted of the fruit being washed under running water and then immersed in a chlorine solution of 150 mg L^{-1} for 10 minutes and then dried at room temperature.

The data presented in Figure 10 on the luminosity of the peel of atemoya fruit allowed us to evaluate its behavior during the storage period, and the interpretation of these results showed the light green color of the fruit during storage; this interpretation is based on the CIELAB system (CIE Subtractive Color System (International Commission for "L'Eclairage") - Lab), in relation to the values found in this experiment. In interpreting the data, there was a 10% reduction between the 18 days of storage (56.6 and 51.4%). This reduction in brightness was characterized by a loss of pigmentation in the peel, which is the first criterion used in the acceptance or rejection of a product by the consumer; the explanation for the loss of luminosity in the experiment may be related to the natural influence of the respiration process during storage, this statement corroborates Chitarra and Chitarra (2005) who report that the variation in fruit color may be related to the temperature and atmospheric conditions in which the fruit is stored as factors of direct influence on its metabolic processes, data found by Torres (2008) showed a loss of rind luminosity in atemoya cv. 'Thompson' on the 6th day of storage under conditions of 22-25°C and 60 ± 10% RH; this difference may be related to the storage method, which differs from this experiment; this statement corroborates the value of this experiment.

Figure 10 - Luminosity values (L*) of the peel of 'Thompson' cv. atemoia, treated with organic acids and packed in expanded polystyrene and polyvinyl chloride trays and stored at 15 ± 0.2°C and 90 ± 2% RH for 18 days. Botucatu SP

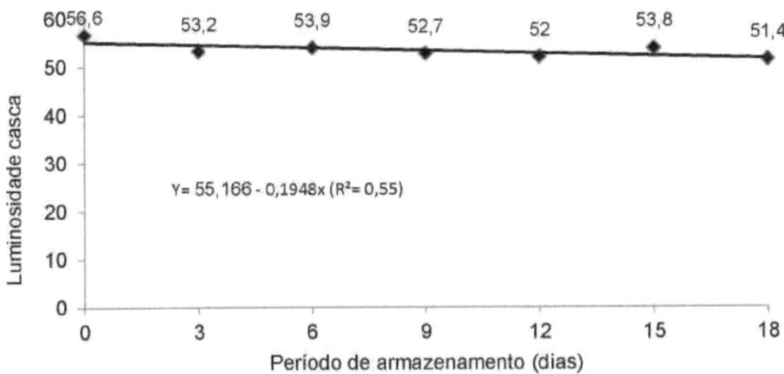

The chromium values found in the peel of this experiment showed no statistical difference between the treatments and their interaction with storage, as shown in Table 9; this statement makes it possible to interpret that the organic acids did not interfere with the variation in the saturation of the peel color during the natural ripening process. The relevance of studying the chroma of the peel is as an analysis factor for characterizing the possible determination of the harvest point in fruit. This statement corroborates Pontes (2004), who states that fruit color is an important quality attribute for consumers, in terms of the negative or positive influence on the perception of sensory attributes, which can help them choose products for consumption.

Table 9 - Peel chroma values of 'Thompson' cv. atemoia, treated with organic acids and packed in expanded polystyrene and polyvinyl chloride trays and stored at 15 ± 0.2°C and 90 ± 2% RH for 18 days Botucatu SP

Treatment	Storage period (days)							Average
	0	3	6	9	12	15	18	
Control *	38,2	34,4	36,4	33,7	37,7	34,6	30,2	35,0
Ascorbic 1%	38,2	38,8	35,6	37,8	37,2	34,1	33,9	36,5
Ascorbic 2%	38,2	37,7	35,7	35,9	38,9	33,4	22,5	34,6
Ascorbic 3%	38,2	37,6	34,9	35,3	36,2	32,7	35,8	35,8
Citrus 1%	38,2	38,1	36,1	34,4	32,2	33,4	32,9	35,1
Citric 2%	38,2	36,0	36,7	36,7	31,6	36,2	33,4	35,5
Citrus 3%	38,2	37,1	36,3	38,1	35,9	29,4	24,8	34,3
C.V. (%)	11,5							

The means of the treatment and interaction were not significant at 5% probability, according to Tukey's test.

(*) The control treatment is characterized by the fruit being washed under running water and then immersed in a chlorine solution of 150 mg L^{-1} for 10 minutes and then dried at room temperature.

The chroma values in the peel found in this experiment, when the storage period was evaluated, showed a reduction of around 20% between day zero and day 18 of storage, respectively (38.2% and 30.5%), as can be seen in (Figure 11); these data make it possible to interpret that the color went from green to yellowish green, possibly as a result of the fruit's natural ripening process; Minolta (1998) mentions that according to the CIELAB system, the lower the average chroma, the lower the intensity of the green or orange colour, and the evaluation of peel chroma is an important item for characterizing the possible existence of anthocyanins in the fruit, which could express changes in colour; This statement corroborates Simón; Hernandez; Gonzalez (2002) who suggest the existence of anthocyanins in fruit, mainly in the peel and occasionally in the pulp; This can be explained by the fact that anthocyanins are phenolic compounds belonging to the flavonoid group, a group of natural pigments widely distributed in the plant kingdom, soluble in water, and which in nature are associated with sugar molecules, being highly unstable at high temperatures, and the final color can be altered depending on factors such as: pH, luminosity, concentration of dissolved anthocyanins, presence of ions, sugars and hormones.

Figure 11 - Chroma values in the peel of 'Thompson' cv. atemoia, treated with organic acids and packed in expanded polystyrene and polyvinyl chloride trays and stored at 15 ± 0.2°C and 90 ± 2% RH for 18 days. Botucatu SP

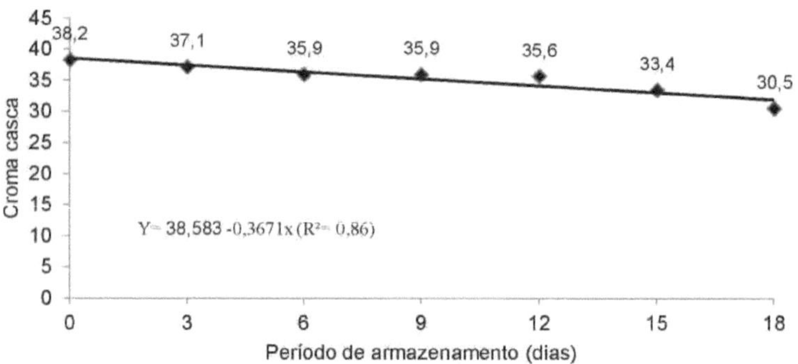

The *°Hue* in the skin of the atemoyas treated with organic acids did not vary statistically between treatments, nor did their interaction with storage, as can be seen in (Table 10). This makes it possible to interpret that the organic acids did not interfere with the color tone during the natural ripening process of the atemoya. The perception of tone is clarified by Ramos; Gomide (2007) who describe the physical action that takes place in the food when it receives a beam of light. When the food receives the beam of light, it absorbs part of the beam and reflects another part. This reflected part is what the human brain characterizes as color and thus enables the interpretation of quality, which influences the consumer's choice of consumption, In addition to this statement, Koblitz (2008) explains that in plant cells we find the ion Mg^{+2} , which is found in chlorophyll and acts on color, but this ion is easily eliminated by reaction with organic acids from the vacuole, resulting in the olive-green color pheophytin;

This statement may explain the non-significance of the treatments in this experiment, since the atemoyas treated with organic acids may have shown the respective loss of Mg+ ion[2] and thus standardized the coloring that occurred with the atemoyas in the experiment.

Table 10 - °*Hue* values in the peel of 'Thompson' cv. atemoia, treated with organic acids and packed in expanded polystyrene and polyvinyl chloride trays and stored at 15 ± 0.2°C and 90 ± 2% RH for 18 days. Botucatu SP

Treatment	Storage period (days)							Average
	0	3	6	9	12	15	18	
Control *	114,2	111,4	111,6	112,0	108,6	111,8	103,1	110,4
Ascorbic 1%	114,2	110,9	111,8	111,8	110,2	109,7	116,8	112,2
Ascorbic 2%	114,2	113,1	110,9	110,0	111,4	108,1	92,5	108,6
Ascorbic 3%	114,2	110,5	111,9	111,7	111,0	108,4	110,0	111,1
Citrus 1%	114,2	112,2	112,8	109,5	107,8	108,3	105,7	109,8
Citric 2%	114,2	112,2	111,1	112,0	108,2	107,1	113,8	111,1
Citrus 3%	114,2	108,9	112,1	109,6	109,8	98,0	94,3	106,7
C.V. (%)	5,9							

The means of treatment and interaction were not significant at 5% probability, according to Tukey's test.

(*) The control treatment is characterized by the fruit being washed in running water and then immersed in chlorine solution at 150 mg L^{-1} for 10 minutes and then dried at room temperature.

The average variation of the ° *Hue* in the skin of taro during the storage period is described in Figure 12, which shows values from 114.2 ° *Hue* for day zero to 105 ° *Hue* for day 18, these data characterize a difference of only 9 ° *Hue* throughout the experiment period and define the color as dark green, Koblitz (2008) characterizes the loss of green color in fruits by the structural decomposition of chlorophyll; this process is caused by several factors, among them: changes in pH, enzyme activity (chlorophyllase) and the presence of oxidizing, enzymatic or chemical systems, data obtained by Silva (2011a) showed variations between ° *Hue* values for atemoya cv. 'Gefner' from 101.89° to 59.75°; 103.72° to 63.65°; 107.62° to 69.68°; 111.52° to 75.68°; from the 1st to the 25th day of storage for doses of 0, 200, 400, 600 ηL.L^{-1} of 1-MCP respectively, this change indicates the evolution of the color of the skin of the atemoia from dark green to light green, this statement reflects the observed por Baten (1990), studying atemoia cv. 'African Pride' stored without packaging at a temperature of 16°C showed deterioration of the external appearance on the 2nd and 6th day; this data corroborates Torres (2008), who evaluated the external appearance of 'Thompson' cv. atemoia stored at 15 ± 1°C and 70-80% RH and characterized external discoloration on the 6th day of storage; these data differ from the experiment carried out here, as it obtained good colouring until the 18th day of storage.

Figure 12 - °*Hue* values in the peel of 'Thompson' cv. atemoia, treated with organic acids and packed in expanded polystyrene and polyvinyl chloride trays and stored at 15 ± 0.2°C and 90 ± 2% RH for 18 days. Botucatu SP

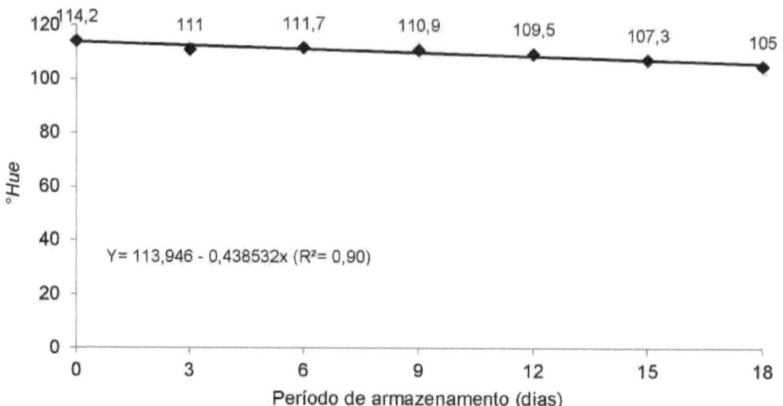

The luminosity values (L*) obtained from the atemoya pulp showed no statistical difference between the treatments with organic acids and the control treatment, nor did they show an interaction between storage time and treatments, as shown in Table 11. This statement demonstrates the non-interference of the use of organic acids in the variation of the luminosity of the atemoya pulp cv. Thompson', but does not exclude the importance of the study, because the characterization of the determination of methodology for controlling the brightness of fruit pulp can be useful in the choice of the same for consumption, by the final consumer, this statement corroborates with Anese; Fronza, (2015) who characterizes the study of fruit color as an auxiliary factor in determining the harvest point; as is sugar, pH, starch and soluble solids.

Table 11 - Luminosity values (L*) in 'Thompson' cv. atemoia, treated with organic acids and packed in expanded polystyrene and polyvinyl chloride trays and stored at 15 ± 0.2°C and 90 ± 2% RH for 18 days. Botucatu SP

Treatment	Storage period (days)							Average
	0	3	6	9	12	15	18	
Control *	96,0	95,3	95,4	94,0	94,6	94,8	94,0	94,9
Ascorbic 1%	96,0	96,6	95,2	93,6	94,6	95,5	91,7	94,7
Ascorbic 2%	96,0	96,3	95,1	91,6	95,3	94,8	87,2	93,7
Ascorbic 3%	96,0	95,3	94,0	94,4	94,2	95,1	92,8	94,5
Citrus 1%	96,0	96,1	96,2	94,8	95,3	93,6	92,8	95,0
Citric 2%	96,0	96,1	95,3	94,0	89,4	94,4	92,8	94,0
Citrus 3%	96,0	96,3	93,3	92,6	93,1	91,2	92,3	93,5
C.V. (%)	2,5							

The means of treatment and interaction were not significant at 5% probability, according to Tukey's test.

(*) The control treatment is characterized by the fruit being washed in running water and then immersed in a chlorine solution of 150 mg L^{-1} for 10 minutes and then dried at room temperature.

The variation in luminosity (L*) obtained in the taro pulp showed values of 96 for day zero and 91.9 for the 18th day of storage, a difference of 4.1 (L*), a negligible value for the period, which demonstrated the stability of the fruit's luminosity during storage; the possible explanation for this behavior may be related to the methodology adopted in the preservation, Minolta (1998) characterizes according to the CIELAB system, the value of (L*), between zero and 100, the value zero being totally black and 100 totally white; This statement can be characterized for the data obtained in this experiment, that the pulp of the atemoia has a white color, a factor of positive influence for a criterion of visual acceptance of the product, this statement corroborates with Lundgren (2017), which cites the differentiation in the luminosity values of the pulp of atemoia cv. 'Thompson' treated with different doses of 1-MCP over 18 days of storage, obtaining results very similar to those of this experiment.

Figure 13 - Luminosity values (L*) of the pulp in 'Thompson' cv. atemoia, treated with organic acids and packed in expanded polystyrene and polyvinyl chloride trays and stored at 15 ± 0.2°C and 90 ± 2% RH for 18 days. Botucatu SP

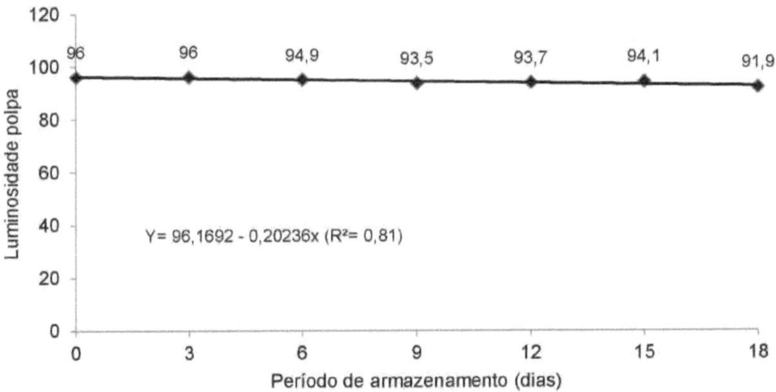

The pulp chroma values found in the atemoya fruit showed that there was no statistical variation between the treatments and their interaction with the storage period, as can be seen in Table 12; This makes it possible to interpret the fact that the organic acids did not act on the natural ripening process of the atemoia fruit in terms of interfering with the saturation of the pulp color, which can be used as an indicator of consumer choice.

Table 12 - Pulp chroma values of atemoia cv. 'Thompson', treated with organic acids and packed in expanded polystyrene and polyvinyl chloride trays and stored at 15 ± 0.2°C and 90 ± 2% RH for 18 days. Botucatu SP

Treatment	Storage period (days)							Average
	0	3	6	9	12	15	18	
Control *	11,0	11,6	12,2	13,3	13,4	12,7	14,7	12,7
Ascorbic 1%	11,0	10,9	11,9	13,1	13,7	12,7	14,3	12,5
Ascorbic 2%	11,0	11,0	11,5	13,5	12,2	13,6	16,3	12,7
Ascorbic 3%	11,0	11,0	12,2	11,9	12,6	13,5	14,8	12,4
Citrus 1%	11,0	11,0	11,2	12,2	12,8	14,0	13,6	12,3
Citric 2%	11,0	13,2	11,9	13,4	14,4	13,3	14,2	13,1

Citrus 3%	11,0	11,1	13,2	13,5	13,9	14,3	14,8	13,1
C.V. (%)	9,6							

The means of treatment and interaction were not significant at 5% probability, according to Tukey's test.

(*) The control treatment is characterized by the fruit being washed under running water and then immersed in a chlorine solution of 150 mg L^{-1} for 10 minutes and then dried at room temperature.

The chroma values of the atemoia pulp varied from 11 on day zero to 14.6 on the 18th day of storage, which characterizes a creamy color; This data is shown in Figure 14. The possible explanation for the variation could be that the chroma changes as a result of the concentration of sugars and carotenoids, which would influence the variation in values due to their degradation; this statement corroborates Montes et al., (2005); Sinnecker et al. (2002) and Meléndez-Martinez et al, (2003), who point out the importance of colorimetry in characterizing the color of different pigments, such as anthocyanins, chlorophyll and carotenoids, as well as assessing the color of food and enabling the product to be valued using visual criteria; data obtained by Torres (2008) on 'Thompson' cv. atemoia can prove that the internal appearance of the fruit was not altered up to the 3rd day of storage, regardless of the treatment used; either at room temperature 22-25.3°C and 60 ± 10% RH or at 15 ± 1°C and 70-80% RH or at 8 ± 1°C and 70-80% RH; these data differ from Lundgren (2017) who observed in 'Thompson' cv. atemoia treated with different doses. 'Thompson' treated with different doses of 1-MCP and stored under refrigeration (15 ± 1°C and 90 ± 5% RH) increased the pulp chroma values for all treatments until the 15th day of storage; these data are similar to those obtained in this experiment.

Figure 14 - Pulp chroma values of atemoia cv. 'Thompson', treated with organic acids and packed in expanded polystyrene and polyvinyl chloride trays and stored at 15 ± 0.2°C and 90 ± 2% RH for 18 days. Botucatu SP

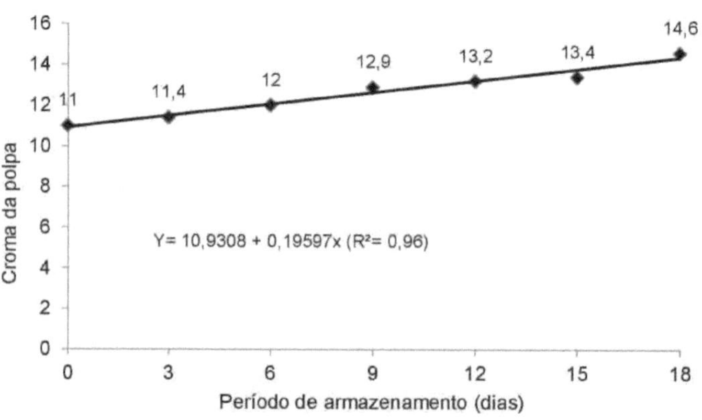

The ° *Hue* is characterized by the hue index, which makes it possible to analyse the difference in color of the fruit during its ripening process. In this experiment, the values did

not show any significance of the treatments or their interaction with the storage period, according to (Table 13). This allows us to interpret that the organic acids did not interfere with the natural hue of the atemoya ripening process.

Table 13 - °*Hue* values in the pulp of 'Thompson' cv. atemoia, treated with organic acids and packed in expanded polystyrene and polyvinyl chloride trays and stored at 15 ± 0.2°C and 90 ± 2% RH for 18 days. Botucatu SP

Treatment	Storage period (days)							Average
	0	3	6	9	12	15	18	
Control *	91,9	91,7	93,1	93,6	94,4	92,4	93,8	93,0
Ascorbic 1%	91,9	92,7	93,4	93,0	93,0	93,3	92,4	92,8
Ascorbic 2%	91,9	92,2	91,9	94,3	93,1	92,0	84,8	91,5
Ascorbic 3%	91,9	92,7	93,8	94,3	93,0	93,0	91,9	92,9
Citrus 1%	91,9	92,9	94,6	94,3	94,2	92,9	91,9	93,2
Citric 2%	91,9	91,4	93,2	94,2	93,3	90,8	88,6	91,2
Citrus 3%	91,9	91,7	92,3	93,2	93,4	92,0	91,9	92,5
C.V. (%)	2,2							

The means of treatment and interaction were not significant at 5% probability, according to Tukey's test.

(*) The control treatment is characterized by the fruit being washed in running water and then immersed in chlorine solution at 150 mg L^{-1} for 10 minutes and then dried at room temperature.

The analysis of the data found in Figure 15 shows that the °*Hue* values ranged from 91.9 on day zero to 90.8 on the 18th day of analysis. This data may characterize the color of white and light pulp; the possible explanation can be made as a function of reflectance (fraction of incident light that was not absorbed or transmitted, but was reflected by the system); Theoretically, a white object has a reflectance curve close to 100% reflectance at visible wavelengths. This statement characterizes °*Hue* as the value in degrees corresponding to the three-dimensional color diagram 0° (red), 90° (yellow), 180° (green) and 270° (blue). C* is represented by the chroma which defines the intensity of the color. The numerical values of a* and b* in chromatology are converted into °*Hue* and chroma which are the variables that best represent the evolution of color during the ripening process. This statement is well-founded according to Guimarâes (2004) because color is defined by three characteristics: hue or tonality, brightness or luminosity and saturation or chroma; hue is the color defined by its wavelength, brightness is the luminosity of the color, referring to its proximity to white or black and chroma is defined as the degree of purity of the color; This statement makes it possible to interpret that during the ripening process the atemoya fruits possibly had the action of the enzyme perodoxidase during their respiration; peroxidase being an enzyme that interferes with color and influences the acceptance of the product by the consumer, when analyzing it by visual criteria only, data obtained by Lundgren (2017) who observed in atemoya cv. 'Thompson' treated with different doses of 1-MCP and stored under refrigeration (15 ± 1°C and 90 ± 5% RH) that the °*Hue* of the fruit pulp showed no significant difference between treatments, days and the treatments x days interaction, i.e. the color of the pulp was not significantly affected by the doses of 1-MCP or the storage period, this statement differs from this experiment.

Figure 15 - °*Hue* values in the pulp of 'Thompson' cv. atemoia, treated with organic acids and packed in expanded polystyrene and polyvinyl chloride trays and stored at 15 ± 0.2°C and 90 ± 2% RH for 18 days. Botucatu SP

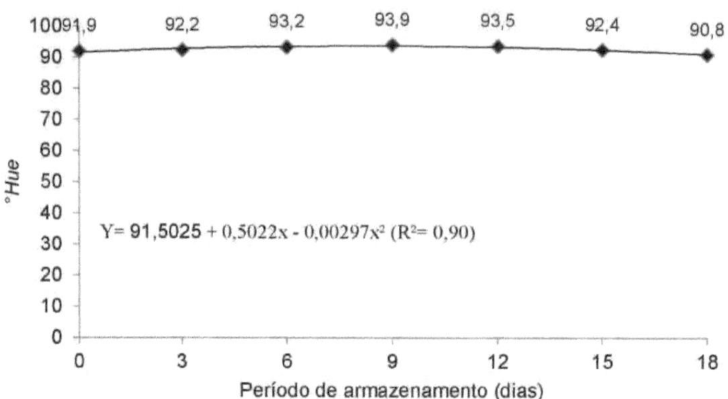

The evaluation of antioxidant activity in this experiment showed that it was most active on the 6th day in the treatments with citric acid at 2% (95.3%) and 3% (94.4%), as shown in Figure 16. The importance of antioxidant activity is related to the fruit's ability to intercept free radicals generated by cellular metabolism, preventing the formation of lesions and loss of cellular integrity, which could subsequently cause the fruit to darken; the amount of organic acids, such as ascorbic and citric acids, can be a determining factor in combating these free radicals, thus maintaining the quality of the fruit after harvest; the data obtained in this experiment corroborates with Leong and Shui (2002) who affirm the interaction of ascorbic acid in the antioxidant action of fruit; while Wang et al. (1996) reported that the percentage contribution of ascorbic acid in the antioxidant action of fruit is less than 15%; Melo et al. (2008) studying the antioxidant capacity of fruit, obtained antioxidant activity of over 70% in pineapple fruit; the explanation for the difference in this experiment may be associated with the extractant, since in the pineapple experiment acetonic extract was used, while in this experiment PA ethyl alcohol was used, Lundgren (2017) studying 'Thompson' atemoia cv. 'Thompson' treated with different doses of 1-MCP and stored under refrigeration (15 ± 1°C and 90 ± 5% RH) found a peak in antioxidant activity on the 6th day of storage, which coincides with the period of this experiment.

Figure 16 - Total antioxidant activity (%) in 'Thompson' cv. atemoia, treated with organic acids and packed in expanded polystyrene and polyvinyl chloride trays and stored at 15 ± 0.2°C and 90 ± 2% RH for 18 days. Botucatu SP

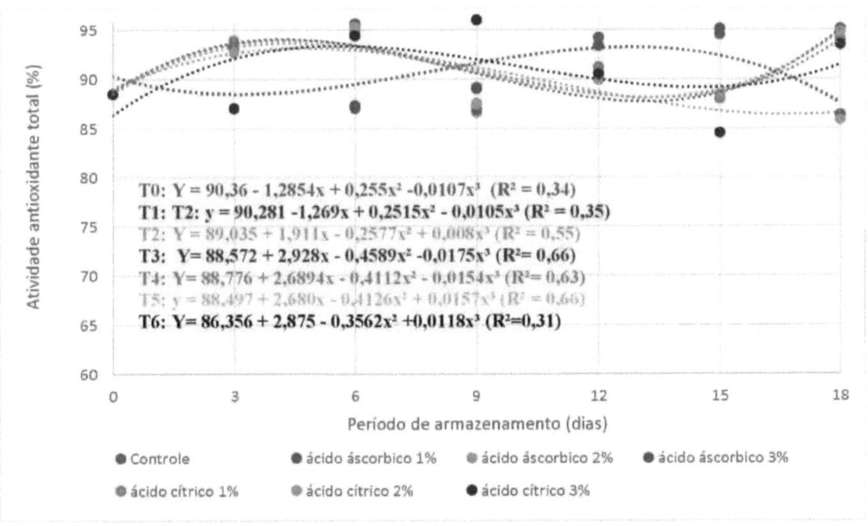

The values found for total phenolic compounds showed that the treatments and their interaction with the storage period were not significant, as can be seen in Table 14. This makes it possible to interpret that the use of organic acids (citric and ascorbic acid) at concentrations of 1%, 2% and 3% did not prove to be efficient or effective in improving the levels of phenolic compounds, which have an antioxidant action in the fight against free radicals resulting from the ripening metabolism of the atemoya.

Table 14 - Values of total phenolic compounds (mg of gallic acid 100 g^{-1} of pulp), in 'Thompson' cv. atemoia, treated with organic acids and packed in expanded polystyrene and polyvinyl chloride trays and stored at 15 ± 0.2°C and 90 ± 2% RH for 18 days. Botucatu SP

Treatment	Storage period (days)							Average
	0	3	6	9	12	15	18	
Control *	672,9	654,2	661,4	1599,4	829,2	1843,0	1000,0	1037,2
Ascorbic 1%	672,9	700,5	583,8	755,5	1037,9	1141,8	552,5	777,8
Ascorbic 2%	672,9	727,3	547,9	823,6	1695,3	861,3	459,2	826,8
Ascorbic 3%	672,9	1036,2	751,5	1977,6	983,8	1066,5	961,4	1064,3
Citrus 1%	672,9	751,2	376,3	733,5	786,1	2011,5	955,6	898,2
Citric 2%	672,9	921,0	881,3	1363,0	1099,5	1246,4	979,9	1023,4
Citrus 3%	672,9	882,7	593,5	551,4	1687,5	822,8	860,4	867,3
C.V. (%)	60,4							

The means of treatment and interaction were not significant at 5% probability, according to Tukey's test.

(*) The control treatment is characterized by the fruit being washed in running water and then immersed in chlorine solution at 150 mg L^{-1} for 10 minutes and then dried at room temperature.

The variations found in the levels of total phenolic compounds found in the atemoias may be related to the different stages of ripeness of each fruit; these showed a behavior that characterized their increase in concentration from the 6th day of storage onwards, 628 mg of gallic acid 100 g^{-1} , and remained increasing until the 12th day of storage, 1159.9 mg of gallic acid 100 g^{-1} , as shown in Figure 17; antioxidant agents such as ascorbic acid and citric acid act in the browning reaction as reducing agents preventing browning; this statement corroborates Marshal, Kim and Wei (2000) who characterize phenols as reducing agents acting to prevent enzymatic browning by reducing o-quinones to diphenols (without color) or reacting irreversibly with them to form stable, colourless products; the characterization of high concentrations of total phenolics allows us to interpret the plant's greater capacity to preserve the lipid oxidation of the tissues and thus preserve the colour; Angelo and Jorge (2007), emphasized that phenolic compounds contribute to the control of fruit pigmentation. They are free radical scavengers and are very efficient at preserving autoxidation, which characterizes them as antioxidants, like flavonoids (anthocyanins and flavonols), phenolic acids (benzoic and cinnamic acids) and coumarins;This statement corroborates Leong and Shui, (2002), who characterize that the quantity and characteristics of phenolic compounds vary depending on the type, variety and degree of ripeness of the fruit as well as the climatic and edaphic conditions of cultivation, these data corroborate with Lundgren (2017), who obtained the highest value of phenolic compounds in atemoya cv. [0]'Thompson' on the 6th day of storage corn 950 mg of gallic acid per 100 g of fresh pulp; confirming the peak concentration on the 6th day of storage, as in this experiment; Roesler et al. (2007) differed in that they found phenolic compounds in araticum fruit, genus Annona, in greater quantity in the peel with 94.19mg 100g^{-1} MS, followed by the seeds, with 53.02mg 100g^{-1} MS and pulp, 10.38mg 100g^{-1} MS; the importance of the value found in this experiment for phenolic compounds is relevant, because in the respiratory process and in various oxidative reactions that occur in the aerobic cells of the fruit, free radicals are formed, which cause damage to the organism; therefore, cells depend on a certain antioxidant capacity to provide protection against the harmful effects of free radicals; This statement corroborates Podsedek (2007), who characterizes the existence of an integrated antioxidant system in the tissues, which consists of an arrangement of various fat-soluble components (vitamin E; carotenoids) and water-soluble ones (ascorbic acid; This statement corroborates Degâspari and Waszczynskyj (2004), who say that red/blue fruits are important sources of phenolic compounds in diets.

Figure 17 - Total phenolic compounds (mg gallic acid 100 g^{-1}), in 'Thompson' cv. atemoia, treated with organic acids, packed in expanded polystyrene and polyvinyl chloride trays and stored at 15 ± 0.2°C and 90 ± 2% RH for 18 days. Botucatu SP

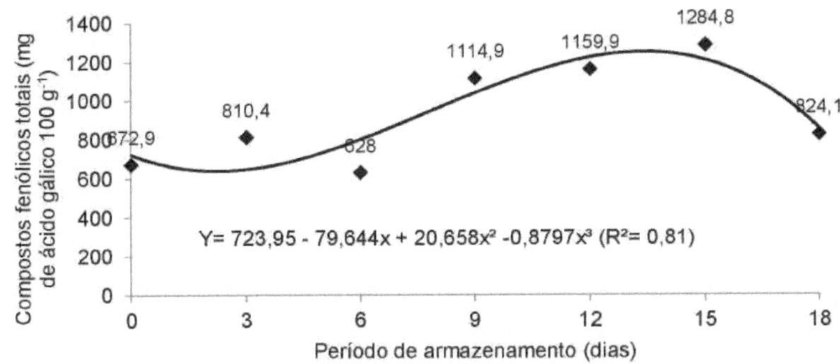

The data obtained on polyphenoloxidation (PPO) showed that there was no significance between the treatments and their interaction with the analysis periods, as shown in Table 15. This allows us to interpret that the use of organic acids did not interfere in the process of controlling the browning of the atemoya during its natural ripening process.

Table 15 - Polyphenoloxidase activity (μmol catechol transformed min^{-1} g^{-1} fresh matter) in 'Thompson' atemoia cv. treated with organic acids and packed in expanded polystyrene and polyvinyl chloride trays and stored at 15 ± 0.2°C and 90 ± 2% RH for 18 days. Botucatu SP

Treatment	Storage period (days)							Average
	0	3	6	9	12	15	18	
Control *	0,404	0,328	0,431	0,167	0,385	0,205	0,284	0,315
Ascorbic 1%	0,404	0,381	0,360	0,323	0,347	0,197	0,547	0,365
Ascorbic 2%	0,404	0,263	0,221	0,252	0,356	0,448	0,472	0,345
Ascorbic 3%	0,404	0,268	0,426	0,202	0,319	0,235	0,403	0,322
Citrus 1%	0,404	0,327	0,287	0,272	0,289	0,200	0,223	0,286
Citric 2%	0,404	0,341	0,344	0,134	0,296	0,293	0,556	0,338
Citrus 3%	0,404	0,198	0,279	0,324	0,380	0,306	0,272	0,309
C.V. (%)	51,9							

The means of treatment and interaction were not significant at 5% probability, according to Tukey's test.

(*) The control treatment is characterized by the fruit being washed in running water and then immersed in chlorine solution at 150 mg L^{-1} for 10 minutes and then dried at room temperature.

The results presented in Figure 18 show the behavior of the enzyme polyphenoloxidase (PPO) in atemoyas during the 18 days of storage, with 0.404 (μmol catechol transformed min g^{-1-1} of fresh matter) on day zero, and on the 9th day of storage there was a reduction in the enzyme's values to 0.239 (μmol catechol transformed min^{-1} g^{-1} of fresh matter). This reduction may be related to the lower oxidation of phenols and o-quinones in the presence of molecular oxygen; Chitarra and Chitarra (2005) state that polyphenol oxidase is the result of the oxidation of phenolic compounds present in fruit, with the initial product of this reaction being quinones, which form dark compounds called melanin. This statement corroborates Costa (2010) who characterizes the importance of the action of ascorbic acid in copper sequestration, reducing quinones to phenols before they form dark pigments; the reduction of PPO values in the fruit is important if they are to have a longer shelf life; its action can be controlled by reducing substances such as ascorbic acid in plant tissue, which prevents browning by inactivating PPO. The data obtained in this experiment corroborates Merodio De La Plaza (1997) who, studying cherimoya, observed a peak in polyphenol oxidase enzyme activity between the 10th and 16th day of analysis after the fruit had been harvested; Torres (2008) emphasized the control of PPO in increasing the shelf life of atemoia cv 'Thompson'. When the fruit was stored at 8 ± 1°C and 70-80% RH, there was very little change in polyphenoloxidase activity during storage, On the 9th day of storage, the fruits kept at this temperature were still immature, with more than 80% green skin color, very resistant to cutting, extremely light and dull flesh and a green fruit aroma. These data disagree with those found by Lundgren (2017), who studied the use of doses of 1- MCP in 'Thompson' atemoia cv. 'Thompson' observed that the levels of polyphenol oxidase were not influenced by the treatments, storage and the interaction between them, and the PPO values found ranged from 0.1 to 0.25 μmol catechol transformed min^{-1} g^{-1} dry matter, values very different from those found in this experiment.

Figure 18 - Polyphenol oxidase activity (μmol catechol transformed min^{-1} g^{-1} of fresh matter), in 'Thompson' cv. atemoia, treated with organic acids packed in expanded polystyrene and polyvinyl chloride trays and stored at 15 ± 0.2°C and 90 ± 2% RH for 18 days. Botucatu SP

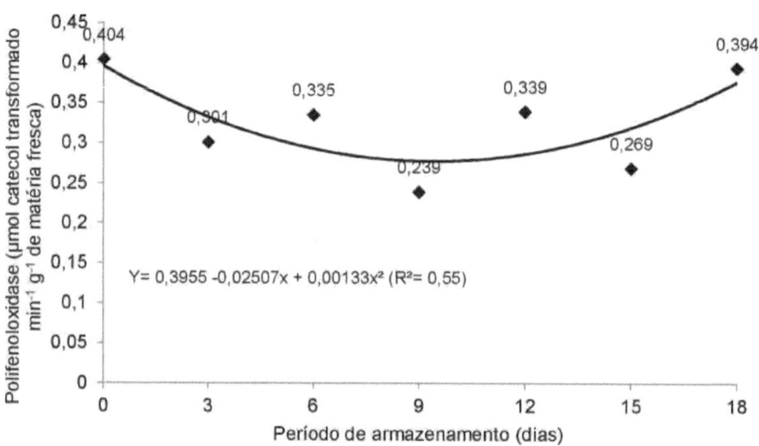

The data obtained with peroxidase (POD) in this experiment showed that the treatments and their interaction with the storage period were not statistically significant, according to the data found in (Table 16), which makes it possible to interpret that the organic acids did not interfere in the natural browning process of the atemoya during the 18 days of storage.

Table 16 - Peroxidase activity (μmol H2O2 transformed min^{-1} g^{-1} fresh matter), in 'Thompson' atemoia cv.

treated with organic acids and packed in expanded polystyrene and polyvinyl chloride trays and stored at 15 ± 0.2°C and 90 ± 2% RH for 18 days. Botucatu SP

Treatment	Storage period (days)							Average
	0	3	6	9	12	15	18	
Control *	0,086	0,056	0,089	0,059	0,086	0,099	0,144	0,088
Ascorbic 1%	0,086	0,089	0,075	0,098	0,128	0,104	0,152	0,104
Ascorbic 2%	0,086	0,073	0,156	0,082	0,310	0,124	0,124	0,136
Ascorbic 3%	0,086	0,063	0,065	0,302	0,217	0,067	0,119	0,131
Citrus 1%	0,086	0,052	0,069	0,130	0,218	0,080	0,220	0,122
Citric 2%	0,086	0,066	0,114	0,125	0,227	0,070	0,088	0,111
Citrus 3%	0,086	0,097	0,094	0,076	0,370	0,095	0,147	0,138
C.V. (%)	87,8							

The means of treatment and interaction were not significant at 5% probability, according to Tukey's test.

(*) The control treatment is characterized by the fruit being washed in running water and then immersed in chlorine solution at 150 mg L^{-1} for 10 minutes and then dried at room temperature.

The behaviour of peroxidase (POD) during the 18-day analysis period showed that the enzyme showed a linear and gradual increase in its activity during this period, with 165% variation between day zero and day 18; data obtained from (Figure 19); This observation can be characterized as the oxidation reaction of phenolic compounds in the presence of hydrogen peroxide, where quinones are obtained as a product, which are unstable and after non-enzymatic oxidation in the presence of o2 polymerize to form melanins, This statement complements Araújo (2008) who cites peroxidase as a thermo-resistant enzyme, so that when it is inactivated, the other enzymes and pathogenic microorganisms will certainly be destroyed, information complemented by Subramanian et al. (1999) who characterize the internal concentration of hydrogen peroxide in plants as small, which limits the activity of the enzyme; the action of POD is therefore more related to slow processes such as the internal browning of fruit; This information is complemented by Zanatta, Zotarelli and Clemente (2006) who found pH 6.3 to be the ideal environment for the action of peroxidase in guava pulp, data from Mosca (2002) analyzing 'Gefner' atemoia cv. 'Gefner' found the highest activity of peroxidase on the 6th day, gradually decreasing until the fruit was fully developed; data that differed from this experiment; data from Torres (2008) showed that the action of peroxidase in guava was influenced by temperature, with storage under conditions of 8 ± 1°C and 70-80% RH being better than 22-25°C and 60 + 10% RH; and 15 ± 1°C and 70-75% RH, this statement corroborates the storage data used in this experiment, data found by Lundgren (2017) showed similar results to this experiment, with atemoya cv. 'Thompson' treated with different doses of 1-MCP and stored under refrigeration at 15 ± 1°C and 90 ± 5% RH; deterioration in fruit caused by enzymatic reactions can occur even at low temperatures, because enzymes are protein catalysts that increase the speed of chemical reactions and are not consumed during the reaction.

Figure 19 - Peroxidase activity (µmol catechol transformed min^{-1} g^{-1} fresh matter), in 'Thompson' atemoia cv. treated with organic acids packed in expanded polystyrene and polyvinyl chloride trays and stored at 15 ± 0.2°C and 90 ± 2% RH for 18 days. Botucatu SP

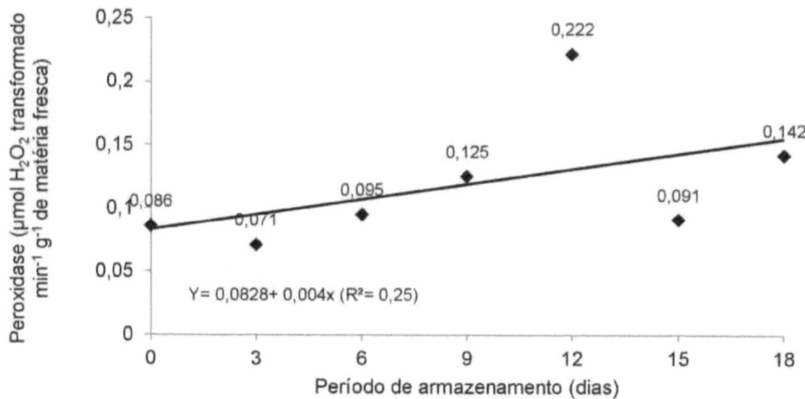

1.4 CONCLUSION

The results obtained in this experiment allow us to conclude that:

- The treatments with citric acid at 1% and 3% showed a climacteric peak in the atemoya on the 6th day with (11.412 and 14.380 mL of CO_2 Kg of $fruit^{-1}$ $hour^{-1}$), respectively 52.10% and 39.65% lower than the control treatment.

- The treatment with 1% ascorbic acid was effective in reducing the mass loss of the atemoia over the 18 days of storage.

- The 2% and 3% citric acid treatments had an effect on the total antioxidant activity of the atemoia up to the 6th day; this was not observed in the other treatments.

- The efficiency of organic acids in terms of producing bioactives capable of interfering in the process of combating free radicals caused by the natural ripening process of the atemoya has not been proven.

REFERENCES

ALI, S.S.; KASOJU, N.; LUTHRA, A. SINGH. A.; SHARANABASAVA, H.; SAHU, A.; BORA, U. Indian medicinal herbs as sources of antioxidants. Food Research International, Toronto, v. 41, p. 1-15, 2008.

ALVES, R.; E.; FILGUEIRAS, H. A. C.; MOSCA, J. L. QUALITY OF NATIVE FRUITS FROM LATIN AMERICA FOR PROCESSING: ATA OR PINHA (*Anonna squamosa* L.). **Frutales:**

Sociedade Tropical de Horticultura, Fortaleza, v. 1, n. 43, p. 77-81, Oct. 2001.

ANESE, Rogério de Oliveira; FRONZA, Diniz. **Post-Harvest Physiology in Fruit Growing**. Santa Maria: Federal University of Santa Maria, 130 p.,2015.

ANGELO, Priscila Milene; JORGE, Neuza. Phenolic compounds in food - a brief review. **Revista Adolfo Lutz**, São Paulo, v. 66, n.1, p. 107-113, 2007.

AOAC. **Official methods of analysis of the Association of Official Analytical Chemistry International**. 13th ed. Washington, 1015 p., 1992.

ARAUJO, Jùlio Maria A. **Food Chemistry:** Theory and Practice. Updated and expanded edition 1ª reprint 5. ed. Viçosa: Editora UFV, 2012. 601 p.

BLEINROTH, E. W.; ZUCHINI, A. G.; POMPEO, R. M. Determination of the physical and mechanical characteristics of avocado varieties and their cold storage. **Coletânea ITAL,** Campinas. v. 7, n. 1, p. 29-81, 1976.

BATTEN. D. J. Effect of temperature on ripening and postharvest life of fruit of atemoya *(Annona cherimola Mill x A. squamosa L.)* cv 'African Pride'. **Science Horticulture**, v 45, p 129-136, 1990.

BRAZIL. Adolfo Lutz Institute (IAL). Normas Analiticas do Instituto Adolfo Lutz - **Métodos Quimicos e Fisicos para Analise de Alimentos**. 4ª edition, Brasilia: 2008.1018p

BRAVERMAN, J. B.S. **Introducción a la bioquimica de los alimentos**. Barcelona: Ômega, 1967, 355 p, chapter 14: vitamins.

BRUINSMA, J.; PAULL, R. E. Respiration during postharvest development of soursop fruit, *Annona muricata* L. **Plant Physiology**, Rockville v. 76, n. 1, p. 131- 138, 1984.

CANO, M. P. *et al* Differences among Spanish and Latin-American banana cultivars: morphological, chemical and sensory characteristics. **Food Chemistry.** n. 59, p. 41119, Amsterdam,1997.

CARDOSO, W. S., PINHEIRO, F. A., PATELLI, T., PEREZ, R., RAMOS, A. M. Determination of sulfite concentration to maintain color quality in dehydrated apples. **Revista Analytica**, n.29, p.66, 2007.

CARVALHO CRL; MANTOVANI DMB; CARVALHO PRN; MORAES RMM. **Chemical analysis of foods:** technical manual. Campinas: Institute of Food Technology. 121p., 1990.

CHITARRA M. I. F.; CHITARRA A. B. **Post-harvest of fruits and vegetables - Physiology and Handling.** Lavras: UFLA. 785p, 2005.

COSTA, A.C.; **Study of the preservation of minimally processed peaches (Prunus persica L.).** Tese (Doutorado), Ciência e Tecnologia Agroindustrial, Universidade Federal de Pelotas, p.77, Pelotas, 2010.

DEGASPARI, Clàudia Helena; WASZCZYNSKYJ, Nina. ANTIOXIDANT PROPERTIES OF PHENOLIC COMPOUNDS. **Visao Acadêmica**, Curitiba, v. 5, n. 1, p. 33-40, jun. 2004, Semestral.

DUARTE-ALMEIDA, J. M. et al. Evaluation of antioxidant activity using β-carotene/linoleic acid system and DPPH radical scavenging method. **Ciência e Tecnologia de Alimentos**, Campinas, v. 26, n. 2, p. 446-452, 2006.

EVANGELISTA, José. **Tecnologia de Alimentos.** 2. ed. Sâo Paulo: Atheneu, 2008. 652 p.

FAGUNDES, A. F.; AYUB, R. A. Physico-chemical characterization of 'Fuyu' cv. persimmons submitted to the application of browning inhibiting agents and stored at 0 °C. **Acta Scientiarum**. Agronomy, Maringa, v. 27, n. 03, p. 403-408, 2005.

FERREIRA, D. F. Sisvar - Analysis of Variance System. Version 5. 3. Lavras MG: UFLA, 2010.

FONTES, Lucina Cristina Brigato. Use of preservative solution and edible films in minimally processed Royal Gala apples: effect on physiology and preservation. 2005. 132 f. Dissertation (Master's Degree) - Science,

Ciência e Tecnologia de Alimentos, Escola Superior de Agricultura "Luiz de Queiroz", Piracicaba, 2005.

GAVA, Altanir Jaime; SILVA, Carlos Alberto Bento da; FRIAS, Jenifer Ribeiro Gava. **Food Technology:** Principles and Applications. Sao Paulo: Nobel, 2009. 511 p.

GONI, O.; SNACHEZ-BALLESTA, M. T.; MERODIO, C.; ESCRIBANO, M. I. Ripeningrelated defense proteins in *Annona* fruit. **Postharvest Biology and Tecnology**, Amsterdam, v. 55, p. 169-173, 2010.

GUIMARAES, Luciano. **Color as information**: the biophysical, linguistic and cultural construction of color symbolism. 3. ed. Sao Paulo: Annablume, 2004.

KOBLITZ, Maria Gabriela Bello. **Food Biochemistry.** Rio de Janeiro: Guanabara Koogan Ltda, 2008. 242 p.

LEONG, L.P.; SHUI, G. An investigation of antioxidant capacity of fruit in Singapore markets. **Food Chem.**, Washington, v. 76, n. 1, p. 69-75, 2002.

LEVINE, M.; PADAYATTY, S.J.; ESPEY, M.G. Vitamin C: A Concentration- Function Approach Yields Pharmacology and Therapeutic Discoveries - a review- **American Society for Nutrition,** v. 2, n. 2, p. 78-88, 2011.

LIMA, G. P. P.; BRASIL, O. G.; OLIVEIRA, A. M. de. Polyamines and peroxidase activity in bean (Phaseolus vulgaris L.) grown under saline stress. **Sci. Agricola**, v. 56, n. 1, 1999.

LIMA, M. A. C. de; ALVES, R. E.; FILGUEIRAS, H. A. C.; ENÉAS-FILHO, J. Respiratory behavior and post-harvest quality of graviola (*Annona muricata L.)* 'Morada' at room temperature. **Revista Brasileira de Fruticultura**, Jaboticabal, v. 25, n.1, p. 49-52, 2003.

LIMA, M. A. C. D.; Alves, R. E.; Filgueiras, H. A. C. Respiratory behavior and softening of graviola (Annona muricata L.) after postharvest treatments with wax and 1-methylcyclopropene. **Ciência e Agrotecnologia**, v. 34, n. 1, p. 155-162, 2010a.

LIMA, Luciana Costa et al. EFFECT OF ASCORBICO ACID ON MINIMALLY PROCESSED "ORANGE FLESH" MELOES. **Alimentaçao e Nutriçao**, Araraquara, v. 2, n. 22, p.291-299, jun. 2011. Quarterly.

LUNDGREN, Giovanna Alencar. **CONSERVATION OF ATEMOIA SUBJECTED TO 1-METHYLCYCLOPROPENE.** 2017. 79 f. Dissertation (Master's Degree) - Agronomic Engineering, Horticulture, Faculty of Agronomic Sciences, Unesp - Botucatu Campus, Botucatu, 2017.

MAIA, G. A.; MESQUITA FILHO, J. A.; BARROSO, M. A. T.; FIGUEIREDO, R. W. Caracteristicas fisicas e quimicas da ata. **Pesquisa Agropecuària Brasileira**, Brasilia, v. 21, n. 10, p. 1073- 1076, 1986.

MARSHALL, M. R.; KIM, J.; WEI, C. Enzymatic browning in fruits, vegetables and seafoods, **Food and Agricultural Organization**, v. 4, n.1, p. 259-312, Washington: FAO, 2000.

MELÉNDEZ-MARTÎNEZ, A. J.; VICARIO, I. M.; HEREDIA, F. J. Application of tristimulus colorimetry to estimate the carotenoids content in ultrafrozen orange juices.

Journal of agricultural and Food Chemistry, v. 51, n. 25, p. 7266-7270, 2003.

MELO, E. A.; MACIEL, M.; LIMA, V. Total phenolic content and antioxidant capacity of frozen fruit pulps. **Alimentos e Nutriçâo**, Araraquara, v. 1, n. 19, p. 6772, mar. 2008. Quarterly.

MENSOR, L. L.; MENESES, F. S; LEITAO, G. G.; REIS, A. S.; SANTOS, T. C.; COUBE, C. S.; LEITAO, S. G Screening of Brazilian Plant Extracts for Antioxidant Activity by the Use of DPPH Free Radical Method, **Phytotherapy Research**, v. 15, n. 2, p. 127-130, 2001.

MERODIO, C.; DE LA PLAZA, J. L. In: MITRA, S. (1997). Postharvest physiology and storage of tropical and subtropical fruits. **CAB International**: Wallingford, 423p.

MINOLTA, K. Precise color communication: quality control from perception to instrumentation. 1998. 59p.

MONTES, C.; VICARIO, I. M.; RAYMUNDO, M.; FEET, R.; HEREDIA, F. J. Application of tristimulus colorimetry to optimize the extraction of anthocyanins from jaboticaba (Myricia jaboticaba Berg). **Food Research International**, v. 38, n. 8-9, p. 983-988, 2005.

MOSCA, José Luiz; CAVALCANTE, Carlos Eliardo Barros; DANTAS, Tatiana Mourâo. **Botanical characteristics of the main anonaceae: Physiological aspects of ripening**. 106. ed. Fortaleza: Embrapa Agroindùstria Tropical, 27 p., 2006.

MOSCA, J. L. **Development, ripening and storage of atemoya (Annona cherimola Mill x Annona squamosa L.) cv. Gefner**. Botucatu: UNESP, 2002. 157 f. Thesis (Doctorate in Agronomy/Horticulture). Botucatu School of Agronomic Sciences - São Paulo State University.

MOSCA, J. L.; LIMA, G. P. P. Respiratory activity of atemoya (*Annona cherimola* Mill. x *Annona squamosa* L.) cv. 'Gefner', during o ripening. Proceedings of The **Interamerican Society For Tropical Horticulture**, Fortaleza, v. 47, n. 1, p. 109110, 2003.

MOTA, R. V.; CORDENUNSI, B. R.; NASCIMENTO, J. R. O.; PURGATTO, E.; ROSSETO, M. R. M.; LAJOLO, F. M. Activity and expression of banana starchphosphorylase during fruit development and ripening. **Planta**, v. 216, n. 2, p. 325-333, 2002.

NASCIMENTO, J. R. O.; VIEIRA JUNIOR, A.; BASSINELO, P.; CORDENUNSI, B. R.; MAINARDI, J. A.; PURGATTO, E.; LAJOLO, F. M. Beta-amylase expression and starch degradation during banana ripening. **Postharv. Biol.Technol**, v. 40, n. 1, p. 4147, 2006.

NELSON, N. A. A photometric adaptation of somogyi method for determination of glucose. Journal Biological Chemistry, Baltimore, v.135, n. 5, p. 375, 1944.

NEVES, C. S. V. J.; YUHARA, E. N. Characterization of the fruits of atemoya cultivars produced in northern Paraná. **Semina: Ciências Agrârias**, Londrina, v. 24, n. 2, p. 311-314, 2009.

NOGUEIRA, E. A.; MELO, N. T. C; MAIA, M. L. Production and commercialization of anonâceas in São Paulo and Brazil. **Informações Econômicas,** v. 35, n. 2, p. 51-54, 2005.

OETTERER, M; REGITANO-D'ARCE, M. A. B.; SPOTO, Marta Helena Fillet. **Fundamentals of Food Science and Technology**. Barueri: Manole, 2010. 612 p. Reprint of the 1st edition[a] - 2010. Chapter 4 Marilia Oetterer and Silene Bruder Silveira Sarmento pages 135 to 195.

OLIVEIRA, M. E. B.; BASTOS, M. S. R.; FEITOSA, T.; BRANCO, M. A. A. C.; SILVA, M. G. G. Avaliação de parâmetros de qualidade fisico-quimicos de pulpas congeladas de acerola, cajà e caju. **Ciências e Tecnologia de Alimentos**, Campinas, v. 19, n. 3, 1999.

PINHEIRO, R. V. R.; MARTELETO, L. O.; SOUZA, A. C. G. de; CASALI, W. D.; CONDÉ, A. R. Productivity and fruit quality of ten varieties of guava, in Visconde do Rio Branco, Minas Gerais, for natural consumption and industrialization. **Revista Ceres**, Viçosa, v. 31, p. 360-387, 1984.

PODSEDEK, A. Natural antioxidants and antioxidant capacity of Brassica vegetables: A review. LWT-Food **Sci. Technol**, v. 40, p. 1-11, 2007.

PONTES, L. V. **Sensory and instrumental evaluation of the color of powdered mixtures for soft drinks, isotonic drinks and gelatin using natural dyes**. Master's dissertation. Postgraduate Program in Food Science and Technology, DTA/Federal University of Viçosa, Viçosa, 2004.

ROCHA, M. C.; SILVA, A. L. B.; ALMEIDA, A.; COLLAD, F. H. Efeito do uso de biofertilizante agrobio sobre as caracteristicas fisico-quimicas na pós-harvest do maracujà-amarelo (*Passiflora edulis f. flavicarpa* Deg.) no município de Taubaté. **Revista Biociências**, Taubaté, v. 7, n. 2, p. 7-13, 2001.

ROESLER. R. et al. Antioxidant activity of cerrado fruits. **Ciência e Tecnologia Alimentos**, Campinas, v. 27, n.1, p. 53-60, 2007.

RAMOS, E. M.; GOMIDE, L. A. M. **Avaliação Objetiva da Cor**. In Avaliaçâo da qualidade de menes: fundamentos e metodologia. Chap 7, ed. UFV, Viçosa p. 287-370, 2007.

RUSSELL, J. B. Another explanation for the toxicity of fermentation acids at low pH: anion accumulation versus uncoupling. Journal of Applied Bacteriology, v. 73, n. 5, p. 363-370, 1992.

SILVA, Ana Veruska Cruz da; MUNIZ, Evandro Neves. QUALITY OF ATEMOYA HARVESTED AT TWO STAGES OF RIPENESS. **Revista Caatinga**, Mossoró, v. 24, n. 4, p. 9-13, 2011.

SILVA, Ana Paula Fioravante Bernardes; LAJOLO, Franco Maria; CORDENUNCI, Beatriz Rosana. EVOLUTION OF STARCH AND SOLUBLE SUGAR CONTENTS DURING THE DEVELOPMENT AND RIPENING OF DIFFERENT MANGO CROPS. **Ciência e Tecnologia de Alimentos,** Campinas, supplement n. 23, p.116-120, dec. 2003.

SIMÓN. G.V., HERNANDEZ. R. M. S., GONZALEZ, M. T. R. Preliminary analysis of anthocyanins in icaco fruit. **Revista Fitotecnia Mexicana**, v. 25, n. 3, p. 261-264, Mexico City, 2002.

SINGLETON, V.L.; ORTHOFER, R.; LAMuELA, R.M. Analysis of total phenols and other oxidation substrates and antioxidants by means of Folin-Ciocalteau reagent. Methods of Enzymology, New York, v.299, p.152-178, 1999.

SINNECKER, P.; GOMES, M. S. O.; ARÊAS, J. A. G.; LANFER-MARQuEZ, u. M. Relationship between color (instrumental and visual) and chlorophyll contents in soybean seeds during ripening. **Journal of Agricultural and Food Chemistry**, v. 50, n. 14, p. 3961-3966, 2002.

SOARES, D.G.; ANDREAZZA, A.C.; SALVADOR, M. Evaluation of compounds with antioxidant activity in Saccharomyces cerevisiae yeast cells. Revista Brasileira de Ciências Farmacêuticas, Araraquara v. 41, n.1, p. 95-100, 2005.

SOMOGYI, M. Determination of blood sugar. **Journal Biologic Chemical**, n. 160, p. 69-73, 1945.

SOUZA, Aline Francieli de; LEÂO, Marcelo Franco. ANALYSES OF THE MOST EFFECTIVE METHODS FOR INHIBITING ENZYMATIC DARKENING IN FRUIT AND VEGETABLES. **Centro Cientifico Conhecer**, Goiânia, v. 8, n. 15, p.117117, 2012.

SOUSA, F. C.; SOUZA, E. P.; CRUZ, C. S. A; GOMES, J. P.; ALMEIDA, F. A. C. PHYSICAL AND PHYSICAL CHEMICAL PARAMETERS OF 'GEFNER' ATEMOIA AT DIFFERENT STAGES OF MATURATION. **Revista Brasileira de Produtos Agroindustriais,** Campina Grande, v. 15, n. 3, p. 329-324, 2013.

SOUZA, Lindomar Maria; CORREIA, Kamila Câmara; SANTOS, Alice Maria Gonçalves dos. **COMPARISON OF METHODOLOGIES FOR ANALYZING pH AND TITULABLE ACIDITY IN HONEY PULP.** 2010. Available at: <https://edisciplinas.usp.br/pluginfile.php/1069903/mod_resource/content/1/Souza 2010.pdf>. Accessed on: 09 Feb. 2018.

SUBRAMANIAN, N.; VENKATESH, P.; GANGULI, S.; SINKAR, V. P. Role of polyphenol oxidase and peroxidase in the generation of blach tea theaflavins. **Journal of Agricultural and Food Chemistry**, California, n. 47, p. 2571-2578, 1999.

TEIXEIRA, Bàrbara de Alencar. **Characterization of the factors involved in the choice and purchase of fruit and vegetables by the adult population of the Federal District.**

2013. 146 f. Dissertation (Master's Degree) - Nutrition Course, Health Sciences, University of Brasilia, Brasilia, 2013.

TORRES, Liz Maria Abi Rached. **POST-HARVEST CONSERVATION OF ATEMOIA cv. 'THOMPSON'.** 2008. 149 f. Dissertation (Master's Degree) - Pharmacy Course, Food Science, Faculty of Pharmaceutical Sciences, Universidade Estadual Paulista "Júlio de Mesquita Filho", Araraquara, 2008.

WANG, H.; CAO, G.; PRIOR, R. L. Total antioxidant capacity of fruits. J. Agric. **Food Chem.**, Washington, v. 44, n. 3, p. 701-705, 1996.

WILEY, R. C. Minimally processed refrigerated fruits and vegetables**. New York: Chapman & Hall, 368 p. 1994**

WONG, D. W. S.; TILLIN, S. J.; HUDSON, J.S.; PAVLATH, A. E. Gas exchange in cut apples with bilayer coatings. **Journal of Agriculture and Food Chemistry**, v.42 n.10, p.2278-2285, 1994.

WOLFE, K.; WU, X.; LIU, R. H. Antioxidant activity of apple peels. **Journal of Agricultural and Food Chemistry**, Easton, v. 51, n. 12, p. 609-614, 2003.

ZANATTA, Caroline Lima; ZOTARELLI, Marta Fernanda; CLEMENTE, Edmar. Peroxidase (POD) and polyphenoloxidase (PPO) in guava pulp (*Psidium guajava* R.). **Ciência e Tecnologia de Alimentos,** Campinas, v. 26, n. 3, p.197-207, 2006.

CHAPTER 2

Use of post-harvest ionizing irradiation on the 'Thompson' atemoya cultivar

SUMMARY

The study of ionizing irradiation in this experiment aims to evaluate its influence on spoilage agents that affect the senescence of 'Thompson' cv. atemoias. The atemoias were irradiated with different doses of cobalt 60 (^{60}Co) and then packed in expanded polystyrene trays (2 fruits/tray) and sealed with 0.020mm polyvinyl chloride films; this was the same as the control treatment, which received no irradiation. The irradiation was carried out at the Centro de Tecnologia das Radiaçôes - CTR IPEN/CNEN - SP, located in the university town of USP (Universidade de Sâo Paulo) in a compact multipurpose irradiator where gamma rays are used. Six doses were used: T_0 = No irradiation (control), T_1 = 0.2kGy, T_2 = 0.4kGy, T_3 = 0.6kGy, T_4 = 0.8kGy, T_5 = 1.0kGy and T_6 = 1.2kGy for each treatment, three replications and two fruits per replication. The treated atemoyas were stored in a cold room at 15 ± 0.2°C and 90 ± 2% RH, and then analyzed every three days for a total of eighteen days. All treatments were analyzed for: skin and pulp chromatology, loss of fresh mass, soluble solids (SS), titratable acidity (TA), pH, ascorbic acid (AA), reducing sugars, respiration rate, specific polyphenoloxidase (PPO) and peroxidase (POD) activity, total antioxidant activity and total phenolic compounds. The experiment used a completely randomized design (DIC) in a 7 x 7 factorial scheme. The data was submitted to analysis of variance using the F test at a 95% probability level. The results showed that the peak of respiration in the atemoia occurred on the 9th day of storage with the 1.2kGy dose respectively (21.853 mL of CO_2 Kg of fruit^{-1} hour^{-1}), three days longer than the control treatment; the differential offered by the producer in terms of the phytosanitary measures adopted in the production, transportation and storage of the atemoia may have reduced the action of contaminants.

Keywords: *Annona x atemoya* Mabb., conservation, cobalt 60, gamma radiation.

2.1 INTRODUCTION

The discovery of the radioactivity of uranium salts was made in 1896 by Antoine Henri Becquerel; in 1963 the US government, through the FDA (Food and Drug Administration), approved the use of irradiation on wheat, wheat flour and bacon, with the aim of inactivating microorganisms that degrade food; in 1964 the use of irradiation was authorized for potatoes and onions, in order to prevent germination. In the 1980s and 1990s, irradiation was regulated in Brazil, especially by the WHO (World Health Organization), FAO (Food and Agriculture Organization of the United Nations) and the International Atomic Energy Agency in 1977, for use in human food. In Brazil, the use of irradiation was regulated by the Ministry of Health, which published Decree No. 72.718 of August 29, 1973, Law No. 7.394 of October 29, 1985 and Resolution RDC No. 21 of January 26, 2001 on irradiated foods (GAVA, 2009).

The aim of using irradiation in this experiment is highlighted by Evangelista, 2008; Chitarra; Chitarra, 2005; O'Beirne, 1989; Fellows, 2006, that the purpose of this technology is to preserve food by protecting it from agents of deterioration, with the aim, for example, of sterilizing or destroying infectious insects in fruit and/or vegetables, not causing an increase in heat, and making it possible to delay the ripening cycle without altering certain organoleptic characteristics of fruit, such as: enhancement of aroma, flavor and color.

Ionizing radiation includes alpha, beta, gamma, X-rays and ultraviolet; what differentiates them is the way they act. Alpha irradiation does not penetrate a sheet of paper, beta irradiation can be interrupted by a sheet of aluminum, and gamma irradiation, used in this experiment, is highly penetrating and can penetrate a block of lead, which is not very thick (GAVA, 2009).

Of the radioactive source elements, cobalt (^{60} Co) is the largest producer of energy; it is an artificial isotope obtained from the irradiation of natural cobalt (EVANGELISTA, 2008).

Radiation doses are quantified in terms of the energy absorbed by the irradiated product and expressed in Gy (gray). A dose of 1Gy is equivalent to the absorption of 1 joule Kg^{-1} of product (CHITARRA; CHITARRA, 2005).

Currently, it is preferred to classify treatments according to the dose used, thus considering low dose (up to 1kGy), medium dose (from 1 to 10kGy) and high dose (from 10 to 50kGy) treatments, with the term radurization consisting of applying doses between 0,4 and 10 kGy of ionizing radiation significantly reduces the microbial load which can influence the reduction of the useful life of the food, such as doses between 0.3 and 1.0 kGy, which have allowed the preservation of mangoes to be extended by one week and bananas by two weeks. The magnitude of this effect depends on the dose and the degree of ripeness reached at the time of treatment (PEREDA, 2007), information that corroborates Chitarra; Chitarra (2005), who cite a minimum dose of between 0.08 and 1.0kGy of radiation necessary to control pathogens or insects in horticultural crops.

The fear of radioactivity has been discussed by the Joint FAO/IAEA/WHO Expert Committee on the Wholesomeness of Irradiated Food (JECFI), which concluded that the maximum average dose of lOkGy "does not represent a toxicological hazard or a special microbiological or nutritional problem in food" (World Health Organization, 1977, 1981)

(FELLOWS, 2006).

Enzymes are resistant to ionizing radiation, which requires the use of high doses of irradiation to inactivate them, which harms food (EVANGELISTA, 2008).

Vitamins are very sensitive to processing, and vitamin C (ascorbic acid) is converted under irradiation into the oxidized form (dehydroascorbic acid), a form of vitamin C (AZEVEDO, 2009).

The study on the use of gamma radiation in the post-harvest preservation of tomatoes concluded that the 0.3kGy dose was the most efficient, while the 1.0kGy dose was not very efficient, but did not compromise the quality of the fruit (VIEITES, 1998).

The use of irradiation on peaches of the cv. 'Biuti' peaches, harvested at the physiological ripening stage, and which received irradiation doses of: 0.0kGy; 0.1 kGy; 0.2kGy; 0.3kGy; 0.4kGy and 0.5kGy through the "GAMMABEAN 650" irradiator, whose source is Cobalt (^{60} Co), showed that the different doses of radiation applied did not promote an increase in the post-harvest shelf life of peaches (CALORE, 2003).

Studies on the effect of packaging and gamma irradiation on the control of microbial contamination of minimally processed mangoes concluded that irradiation was highly effective in controlling contamination by molds and yeasts, since no contamination was found in the fruit treated with a dose of 0.5kGy (VIEITES et al., 2004).

Research using ionizing irradiation found that gamma irradiation with^{60} CO was effective in controlling the ripening of 'Tropic Beauty' peaches, with doses of 0.4 and 0.8kGy being the most suitable (COSTA, 2008).

The use of gamma radiation in the post-harvest preservation of the nectarine cv. 'Sunred', at a dose of 0.4kGy, showed the best visual aspect, the lowest fresh mass losses and the highest pulp firmness, after 28 days of storage in a cold room at 0°C and 90-95% RH (NEVES, 2009).

Research carried out with the aim of verifying the effect of gamma irradiation on the loss of mass and respiratory activity of avocado fruit of the 'Hass' variety; these were evaluated at intervals of: 0, 7, 14, 21, 28 and 35 days, and kept under refrigeration at a temperature of ± 10°C and 90 ± 5% RH; showed that the doses of 0.2 and 0.6kGy were the most efficient in preserving the fruit (DAIUTO, 2010).

In order to characterize the behaviour of gamma radiation in the post-harvest preservation of 'Sunred' nectarines, they were treated with doses of: 0.2kGy, 0.4kGy, 0.6kGy and 0.8kGy, and then stored in a cold room at temperatures of 0°C and 90-95% RH for 28 days. Analyses were carried out every seven days, determining the visual aspect of the fruit, loss of fresh mass, flesh firmness, total titratable acidity (TA), total soluble solids (SS) and the SS/TA ratio. After 28 days of storage, it was found that the fruit subjected to the 0.4kGy dose had the best visual appearance, the lowest loss of fresh mass and the highest pulp firmness, while there were no significant variations in the levels of total titratable acidity, total soluble solids and the SS/AT ratio (NEVES, 2009).

Peaches of the 'Tropic Beauty' variety subjected to ionizing irradiation of 0.4, 0.6, 0.8 and 1.0 kGy respectively, stored in B.O.D. at 0°C and 90 ± 5% RH for a period of 25 days, with evaluations carried out every five days. at 0°C and 90 ± 5% RH for a period of 25 days, with evaluations carried out every five days, it was concluded that gamma irradiation influenced the loss of fresh mass, firmness, pH, titratable acidity, soluble solids, SS/AT ratio, ascorbic acid, reducing sugars and total reducing sugars. The 0.4 and 0.8kGy doses were more suitable for inhibiting the onset of ripening (COSTA, 2008).

The application of 0.3kGy and 0.6kGy doses of gamma irradiation *to fresh* 'Kioto' persimmons proved to be adequate and promoted preservation for 21 days (MENDONÇA, 2016).

The aim of this work was to evaluate the influence of ionizing irradiation on the production of bioactives in atemoya fruit, favoring post-harvest preservation.

2.2 MATERIAL AND METHOD

The "Thompson" cv. atemoias were treated using ionizing irradiation (^{60} Co), with the following doses: T0=No irradiation (control), TI=0.2kGy, T2=0.4kGy, T3=0.6kGy, T4=0.8kGy, T5=1.0kGy and T6= 1.2kGy, for each treatment three replicates were assigned with two fruits per replicate. The atemoias were packed in expanded polystyrene trays and sealed with 0.020 mm polyvinyl chloride film (Photo 3), then placed in Styrofoam boxes measuring 89 cm x 42 cm x 25.5 cm, with a thickness of 3 cm (Photo 4) and transported in a private car with air conditioning, to IPEN - Instituto de Pesquisa Energéticas e Nucleares, located in the Centro de Tecnologia das Radiaçôes - CTR IPEN/CNEN - SP, located in the university city of USP (Universidade de Sâo Paulo), to be subjected to the irradiation doses with^{60} CO in the compact multipurpose irradiator where the Gamma Ray is used (Photo 5).

The atemoias were then returned to the post-harvest laboratory at the Faculty of Agronomic Sciences - FCA - "Jûlio de Mesquita Filho" UNESP - Campus of Botucatu SP, and stored in a cold room at 15 ± 0.2°C and 90 ± 2% RH for 18 days.

During the storage period, laboratory analyses were carried out every three days (0, 3, 6, 9, 12, 15 and 18), following the methodologies described in items 1.2.1 to 1.11.2 of Chapter 1 (Use of organic acids (ascorbic and citric) in the post-harvest period of the 'Thompson' cultivar of atemoya).

2.2.1 Sorting, transporting and receiving the cranberries

The origin, selection and transportation of the fruit complied with the same conditions described in section 1.2 of chapter 1 (Use of organic acids (ascorbic and citric) in the post-harvest of the 'Thompson' cultivar of atemoya).

Photo 3 - 'Thompson' seedlings, packed in expanded polystyrene trays sealed with 0.020mm polyvinyl chloride film for subsequent ionizing irradiation treatment.

Photo 4 - Atemoias cv. 'Thompson', packed for subsequent ionizing irradiation treatment at IPEN.

Photo 5 - Multipurpose irradiator with^{60} CO.

2.3 RESULTS AND DISCUSSION

The data obtained on the respiration of the atemoyas showed that the excess of the treatments with 1.0 kGy and 1.2 kGy, that the respiration peak occurred on the 6th day of storage for the treatment: control (21.459 mL of CO_2 Kg of fruit^{-1} hour^{-1}), 0.4kGy (24.959 mL of CO_2 Kg of fruit^{-1} hour^{-1}), 0.6kGy (26.432 mL of CO_2 Kg of fruit^{-1} hour^{-1}) and 0.8kGy (28.883 mL of CO_2 Kg of fruit^{-1} hour^{-1}), the treatment of 1,0kGy and 1.2kGy treatment showed peak respiration on the 9th day respectively (33.388 mL of CO_2 Kg of fruit^{-1} hour^{-1}) and (21.853 mL of CO_2 Kg of fruit^{-1} hour^{-1}), data shown in (Figure 20). The reduction in respiration may be due to the control of microorganisms which can cause deterioration of the fruit, speeding up its ripening process, a concept which can be explained by radapertization. This statement corroborates Ventura et al. (2010) who characterize ionizing radiation as a method which interacts with the molecules in the fruit in such a way as to create negative and positive ions and thus transfer energy through electrons. The formation of these charges results in chemical and biological effects that prevent cell division in bacteria by disrupting their molecular structure. The energy levels used to achieve this effect are not sufficient to induce radioactivity in food; this statement coincides with Tezotto et al. (2014), who explain the action of gamma radiation as acting to control microorganisms through damage to their DNA. Lethal damage to DNA is spread through cells during the process of cell division, and those that are unable to repair such damage die; Mendonça (2016), studying the use of gamma radiation on 'Kioto' persimmons, concluded that lower doses (0.3 and 0.6 kGy) were more appropriate, prolonging preservation for 21 days, and also found that the technique can be used to improve the preservation of fruit in post-harvest; these data corroborate Neves (2009), who studied the application of gamma radiation to 'Sunred' nectarines. 'Sunred' observed that the dose of 0.4 kGy enabled the fruit to be preserved for 28 days; these statements characterize the use of ionizing radiation, such as that used in this experiment, as a viable technique for post-harvest preservation. No data on the use of gamma radiation on atemoyas was found in the literature.

Figure 20 - Respiratory activity (mL of CO_2 Kg of fruit^{-1} hour^{-1}) as a function of irradiation doses (kGy) in 'Thompson' cv. atemoia and packed in expanded polystyrene and polyvinyl chloride trays and stored at 15 ± 0.2°C and 90 ± 2% RH for 18 days. Botucatu SP

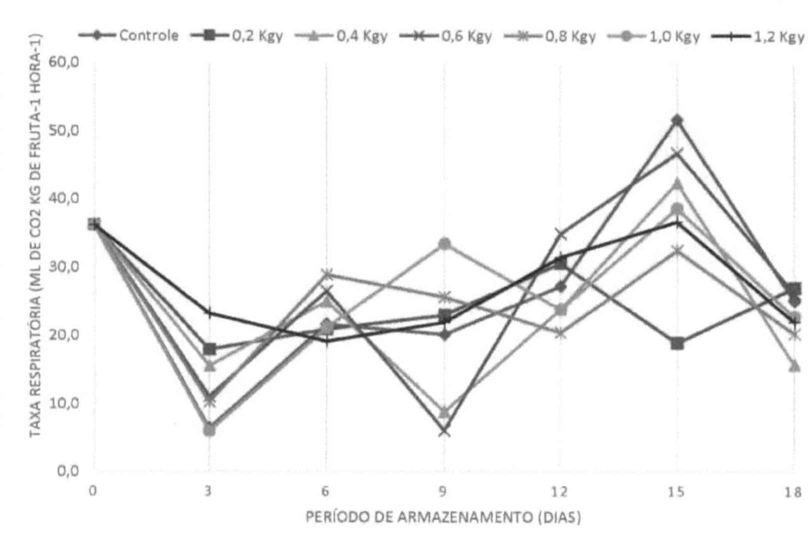

Soluble solids (SS) in irradiated atemoia

The analysis in Figure 21 shows the results of the averages obtained with the doses of irradiation during the storage period. The fruit irradiated with 1.2KGy had average soluble solids of 16.75 °Brix, 12.4% lower than the value found in the control treatment.

The behavior of the soluble solids values obtained in the post-harvest period can be explained by Chitarra and Chitarra (2005), who report that the soluble solids content varies upwards during storage as a result of ripening due to the degradation of polysaccharides; the data obtained in this experiment differ from Lundgren (2017), who found soluble solids values in atemoya cv. 'Thompson' of 32° Brix for all the fruit in all the treatments carried out with 1-MCP and stored under refrigeration (15 ± 1°C and 90 ± 5% RH) for 18 days. This information corroborates Silva (2011) who, studying the quality attributes of 'Gefner' atemoia harvested at two stages of ripeness ("de vez" and ripe), obtained an average soluble solids of 30.47 °Brix.

Figure 21 - Average soluble solids content (°Brix) as a function of irradiation doses (kGy) in 'Thompson' atemoia and packed in expanded polystyrene and polyvinyl chloride trays and stored at 15 ± 0.2°C and 90 ± 2% RH for 18 days. Botucatu SP

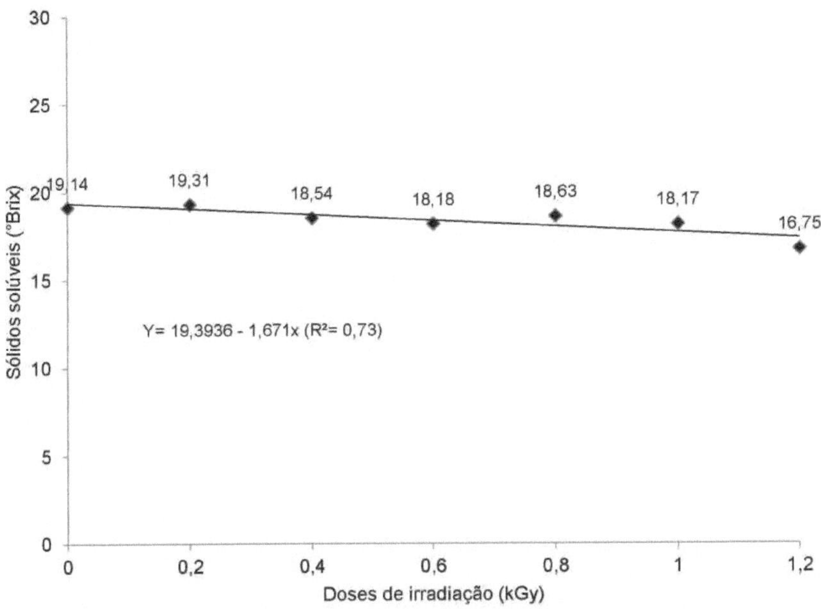

The average soluble solids levels found in the atemoias in relation to storage time showed an increase over the 18-day storage period, with soluble solids levels starting at 13.67 °Brix on day zero and reaching 21.31 °Brix on the 18th day of storage, as shown in Figure 22, These statements make it possible to interpret the possible control of microorganisms that could act in the deterioration of the fruit by interfering with its respiration, as for biochemical criteria, related to the secondary metabolism of the fruit, which includes phenolic compounds, tannins, examples of fruit defense, the data obtained in this experiment corroborate with Souza (2013), studying atemoias cv. 'Gefner' at different stages of ripeness, observed soluble solids concentrations of 20.68° Brix in semi-ripe fruit and 25.95 °Brix for ripe fruit. These data differ from Silva (2011) who, studying the quality of 'Gefner' atemoyas harvested at two stages of ripeness, obtained a soluble solids concentration of 29.7° Brix for 'once' fruit and 31.25 °Brix for ripe fruit, values 71.7% and 68.16% higher than those found in this experiment, for the highest soluble solids content found in 'Thompson' atemoyas at 18 days of storage.

Figure 22 - Average soluble solids content (°Brix) in 'Thompson' cv. atemoia as a function of storage period and packed in expanded polystyrene and polyvinyl chloride trays and stored at 15 ± 0.2°C and 90 ± 2% RH for 18 days. Botucatu

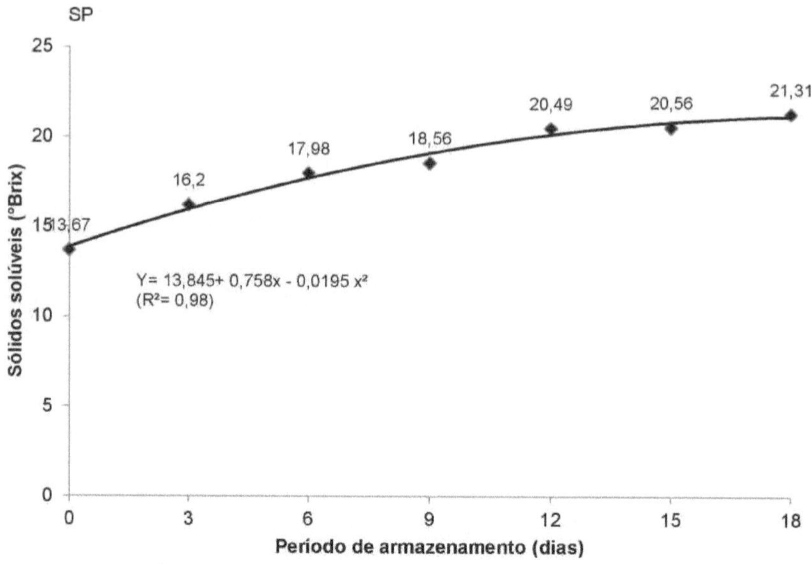

Hydrogen potential (pH) in irradiated atemoia

The study of the influence of gamma radiation on taro for pH values showed through the data obtained in Table 17, the characterization of the non-significance of the doses and their non-interaction with the storage period; This information makes it possible to interpret that the use of the methodology did not interfere with the pH values, which are relevant to the fruit ripening process, as they are one of its indicators, where the increase and decrease in titratable acidity are indicative of fruit ripening.

Table 17 - pH values as a function of irradiation doses (kGy) in 'Thompson' cv. atemoia packed in expanded polystyrene and polyvinyl chloride trays and stored at 15 ± 0.2°C and 90 ± 2% RH for 18 days. Botucatu SP

Doses	Storage period (days)							Average
	0	3	6	9	12	15	18	
Control*	6,06	5,83	5,75	5,91	5,68	5,53	5,83	5,80
0.2kGy	6,06	5,98	6,08	5,68	5,81	5,51	5,81	5,85
0.4kGy	6,06	6,16	5,94	5,80	5,62	5,52	5,78	5,84
0.6kGy	6,06	6,28	5,86	5,87	5,95	5,65	5,81	5,93
0.8kGy	6,06	5,97	5,90	5,89	5,71	5,63	5,85	5,86
1.0kGy	6,06	6,23	5,98	5,75	5,73	5,60	5,86	5,89
1.2kGy	6,06	6,25	5,84	5,74	5,85	5,85	5,68	5,89
C.V. (%)	3,2							

The means of the treatment and interaction were not significant at 5% probability, according to Tukey's test.

(*) The control treatment is characterized by the fruit being washed in running water and then dried at room temperature.

Figure 23 shows the variations in pH values over the course of storage, from 6.06 on day zero to 5.8 on day 18, which represents a percentage decrease of 4.97%. The influence of pH on fruit biochemistry is related to fruit ripening, which coincides with an increase in pH and a reduction in titratable acidity; This information makes it possible to interpret that ionizing irradiation contributed to a decrease in pH during this time, which slowed down the ripening process. This statement corroborates Oh et al. (2006) who reported that irradiation influences pH in such a way that it can decrease or increase with the dose of irradiation due to its interaction with sucrose, glucose and fructose molecules, fragmenting them to form acids. This statement is confirmed by Cardello and Cardello (1998), who assessed the ripening of mango cv. 'Haden' in comparison with the variation in pH over 14 days, found that there was a decrease in pH and an increase in titratable acidity during this period. The data obtained with atemoia in this experiment corroborates Souza (2013) who, when assessing the chemical parameters of atemoia cv. 'Gefner' observed a pH value of 5.36 in the fully ripe fruit; this data confirms Silva (2011), who evaluated the pH value of the 'Gefner' atemoia cv. at pH 5.58; it also corroborates Torres (2008), who, evaluating the post-harvest period of the 'Thompson' atemoia cv. 'Thompson', found pH values of 5.5 in ripe fruit stored at room temperature (22-25°C and 60 + 10% RH) and at 15 + 1°C and 70-75% RH; chemical analysis studies on graviola pulp (a type of anonacea) carried out by Fonseca (2012) found pH values of between 3.55 and 3.45 for ripe fruit, values which differ from this experiment, possibly due to the different nature of the fruit studied.

Figure 23 - pH values in 'Thompson' cv. atemoia as a function of irradiation doses (kGy) and packed in expanded polystyrene and polyvinyl chloride trays and stored at 15 ± 0.2°C and 90 ± 2% RH for 18 days. Botucatu SP

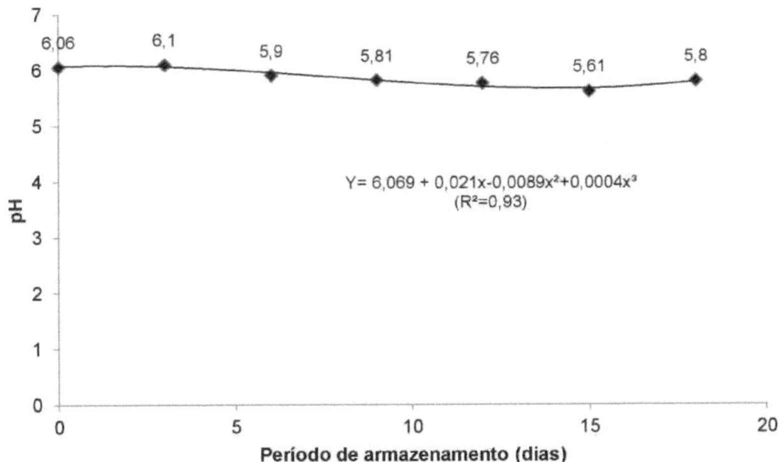

Titratable acidity (TA) in irradiated atemoia

The data obtained on titratable acidity (TA) shows that there was a statistical difference in the IkGy dose during the storage period. The TA levels increased until the 9th day (0.277 g citric acid 100 g^{-1}), and after this period there was a reduction in the results, with 0.093 g citric acid 100 g^{-1} being obtained on the 18th day. The other treatments showed no statistical difference during the storage period, as shown in Figure 24.

This observation makes it possible to interpret that the 1.0kGy treatment was the treatment that acted to control the ripening of the 'Thompson' cv. atemoia, up to the 9th day, thus making it possible for the fruit to last longer in post-harvest. This statement can be clarified by the importance of maintaining high titratable acidity values, as this tends to decrease during the fruit's ripening process.

Data found by Mendonça (2016) studying the influence of ionizing irradiation on the titratable acidity of kaki cv. 'Kioto', obtained a reduction in this up to 21 days, Calore (2003) found the highest levels of titratable acidity in peaches cv. 'Buiuti' irradiated with doses of 0.2kGy and 0.4kGy allowing the fruit to be preserved for 14 days.

Figure 24 - Titratable acidity content (g citric acid 100 g^{-1}) as a function of irradiation dose (kGy) in 'Thompson' cv. atemoia packed in expanded polystyrene and polyvinyl chloride trays and stored at 15 ± 0.2°C and 90 ± 2% RH for 18 days. Botucatu SP

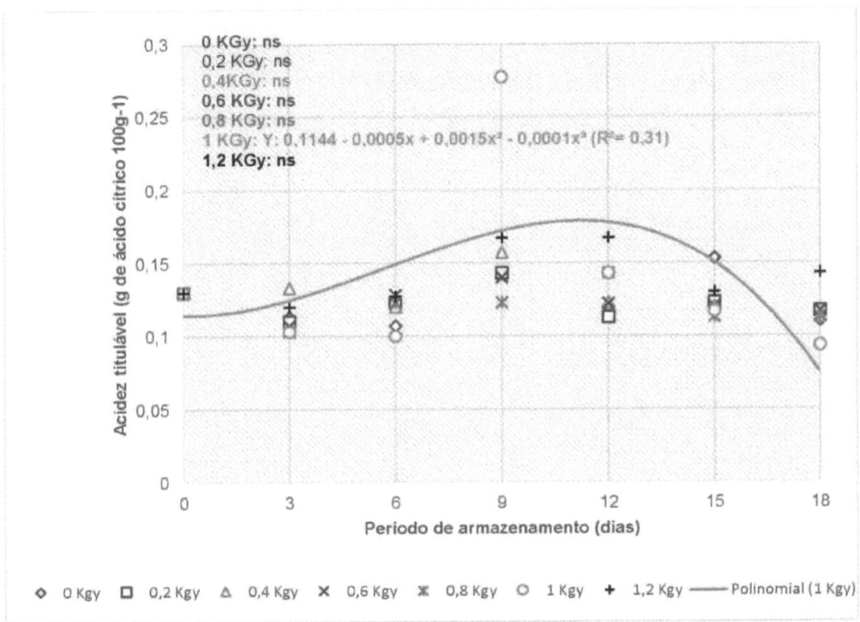

Mass loss (MP) in irradiated atemoia

The analysis data for mass loss (%) is shown in Figure 25. For mass loss (MP) there was a difference between the averages of the treatments (irradiation doses), the lowest MP was observed in the fruit treated with 0.4kGy (0.619%) and the highest loss in the treatment with 1kGy (1.12%). These data allow the interpretation that high doses of irradiation do not favor the control of mass loss in 'Thompson' atemoyas. This information can be complemented by the statement that the loss of fruit mass is mainly associated with the loss of water caused by both transpiration and respiration of the fruit during the ripening process. Calore (2003), studying the preservation of 'Buiuti' peaches, observed that the lowest loss of fresh mass was obtained with a dose of 0.1kGy.

Figure 25 - Average values of mass loss (%) as a function of irradiation doses (kGy) in 'Thompson' atemoia cv. packed in expanded polystyrene and polyvinyl chloride trays and stored at 15 ± 0.2°C and 90 ± 2% RH for 18 days. Botucatu

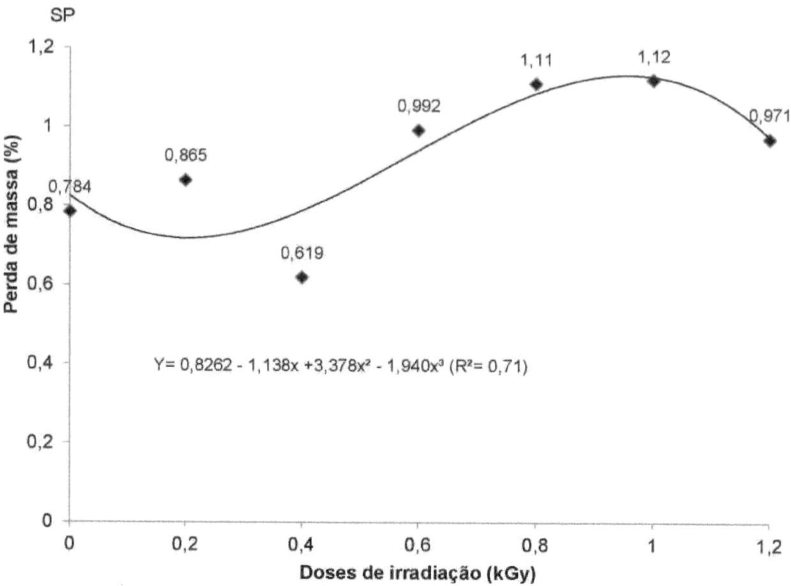

The loss of mass (%) obtained in the irradiated fruit with the relationship between doses and storage time showed constant losses over the course of eighteen days of analysis; this can be seen in Figure 26, which shows losses between the 3rd and 18th day of (0.33%) and (1.781%) respectively, a difference of 539%.

This statement can be explained by the fact that the increase in mass loss in fruit is due to the fact that they lose water and solutes through transpiration and respiration respectively; Chitarra and Chitarra (2005) cite that 3% mass loss can cause a reduction in fruit quality, data obtained by Torres (2008), observed 12% mass loss on the 9th day of analysis in atemoya cv. 'Thompson' stored at 15 ± 1°C and 70-75% RH; without packaging control; these differ from Lundgren (2017), who obtained losses close to 5% in atemoya cv. 'Thompson' on day 9º of storage, in the control treatment, under refrigeration (15 ± 1°C and 90 ± 5% RH), these data differ from this experiment which obtained a loss of 0.951% for the same period; The observations described here corroborate this experiment, as they emphasize the importance of assessing mass loss as an analytical indicator for evaluating a fruit's shelf life, which is inversely proportional to post-harvest losses and directly proportional to the producer's financial gain.

Figure 26 - Mass loss (%) in 'Thompson' atemoia cv. as a result of storage period and irradiation doses (kGy), packed in expanded polystyrene and polyvinyl chloride trays and stored at 15 ± 0.2°C and 90 ± 2% RH for 18 days. Botucatu SP

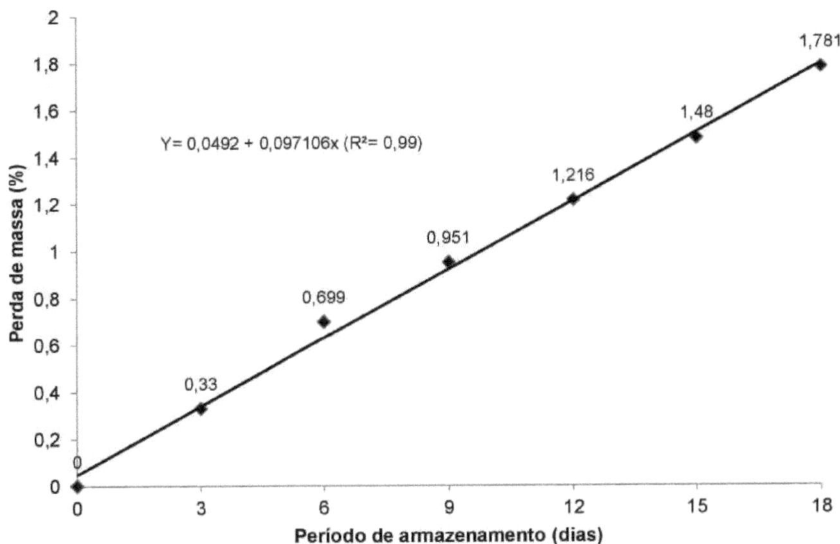

Ascorbic acid in irradiated atemoia

In the analysis of Table 18, the ascorbic acid contents show that the averages of the treatments and their interaction with the storage period in the irradiated atemoyas were not significant at 5% probability, according to Tukey's test; this makes it possible to interpret that the use of ionizing irradiation did not influence the ascorbic acid contents (an antioxidant) during the natural ripening of the atemoyas.

Table 18 - Ascorbic acid content (mg ascorbic acid $100g^{-1}$) in atemoia cv. 'Thompson' as a function of irradiation doses (kGy) packed in expanded polystyrene and polyvinyl chloride trays and stored at 15 ± 0.2°C and 90 ± 2% RH for 18 days. Botucatu SP

Doses	Storage period (days)							Average
	0	3	6	9	12	15	18	
Control*	50,16	41,28	37,92	31,33	33,62	22,46	20,95	33,96
0.2 kGy	50,16	39,33	25,55	19,30	16,54	25,30	10,96	26,73
0.4 kGy	50,16	40,15	39,19	22,99	19,54	20,23	10,17	28,91
0.6 kGy	50,16	41,04	30,92	23,73	33,68	17,54	10,82	29,70
0.8 kGy	50,16	42,59	28,29	21,18	49,20	12,39	9,34	30,45
1.0 kGy	50,16	33,71	34,14	24,66	19,59	10,66	16,61	27,08
1.2 kGy	50,16	41,81	40,95	33,29	34,27	10,98	15,55	32,43
C.V. (%)	29,9							

The means of treatment and interaction were not significant at 5% probability, according to Tukey's test.

(*) The control treatment is characterized by the fruit being washed under running water and then dried at room temperature.

The levels of ascorbic acid found in the atemoyas during the storage period are shown in Figure 27, which shows an increasing decrease over the 18 days of storage. This makes it possible to interpret that during ripening the fruit shows a decrease in ascorbic acid levels, which has a delaying action on the ripening process as it acts in the detoxification of hydrogen peroxide; this statement corroborates Levine et al, (2011) who cite vitamin C acting as an enzyme cofactor, taking part in oxidation-reduction processes, increasing iron absorption and inactivating free radicals; this statement is complemented by Chitarra and Chitarra (2005), who report the influence of ionizing radiation on the chemical change in fruit, in terms of the possible loss of free radical scavengers, such as ascorbic acid, Ventura et al. (2010) explains that ascorbic acid in solution is completely unstable to radiation, but in fruit and vegetables it appears to be completely stable at low treatment doses. Koblitz (2008) characterizes the oxidized product of vitamin C, L-dehydroascorbic acid, as having the same biological activity, but it is not very stable, Its formation is practically instantaneous at alkaline pH, rapid around neutrality and slow in acidic conditions. These statements may explain the non-significance of the treatment means and the non-interaction at 5% probability, by Tukey's test, due to the pH variation found in this experiment.

Figure 27 - Ascorbic acid content (mg ascorbic acid $100g^{-1}$) in atemoya cv. 'Thompson' as a function of irradiation doses (kGy) and packed in expanded polystyrene and polyvinyl chloride trays and stored at 15 ± 0.2°C and 90 ± 2% RH for 18 days. Botucatu SP

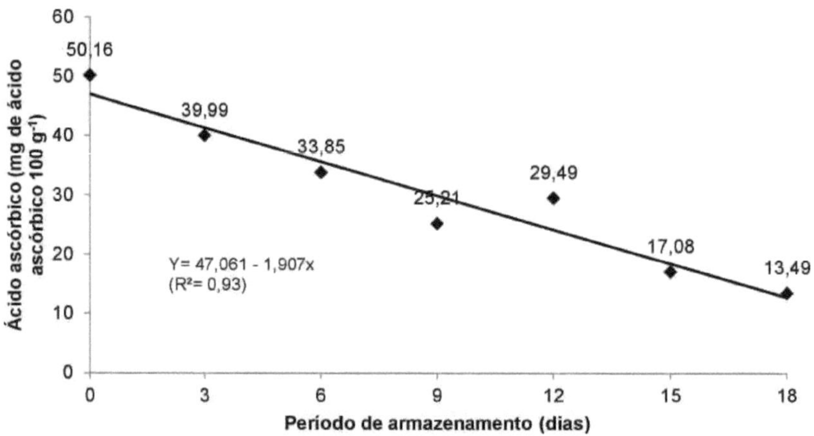

Reducing sugar in irradiated atemoia

The reducing sugar content found in the experiment with ionizing irradiation in the atemoyas showed, in Figure 28, that the treatment with a dose of 0.8kGy presented 4.18% reducing sugar on day zero and on the 18th day of storage the index was 9.16%, a 219% increase over that period, while the treatment with 1,2kGy treatment showed the lowest levels of reducing sugar over the same period. These data indicate that there was a concentration of glucose and fructose, which are sweeter monosaccharides than sucrose (disaccharide -

fructose and a-glucose), and the hydrolysis of sucrose is influenced by pH (very acidic), temperature (low temperature) and time, the non-hydrolysis of reducing sugars kept the fruit as unripe, this statement corroborates Ribeiro and Seravalli (2007) who report that most of the disaccharides found in food are reducing, with the exception of sucrose, which is sensitive to acid hydrolysis and, under its action, is subsequently transformed into D-glucose and D-fructose, which is an indicator of ripeness. This statement corroborates Lundgren (2017), who obtained an increase in the values of reducing sugars during the eighteen-day storage period for 'Thompson' atemoyas treated with different doses. 'Thompson' treated with different doses of 1-MCP, but with higher indices than in this experiment, i.e. on day zero 5% and on day eighteen 20.84%.

Figure 28 - Reducing sugar content (%) in atemoia cv. 'Thompson' as a function of irradiation doses (kGy) and packed in expanded polystyrene and polyvinyl chloride trays and stored at 15 ± 0.2°C and 90 ± 2% RH for 18 days. Botucatu SP

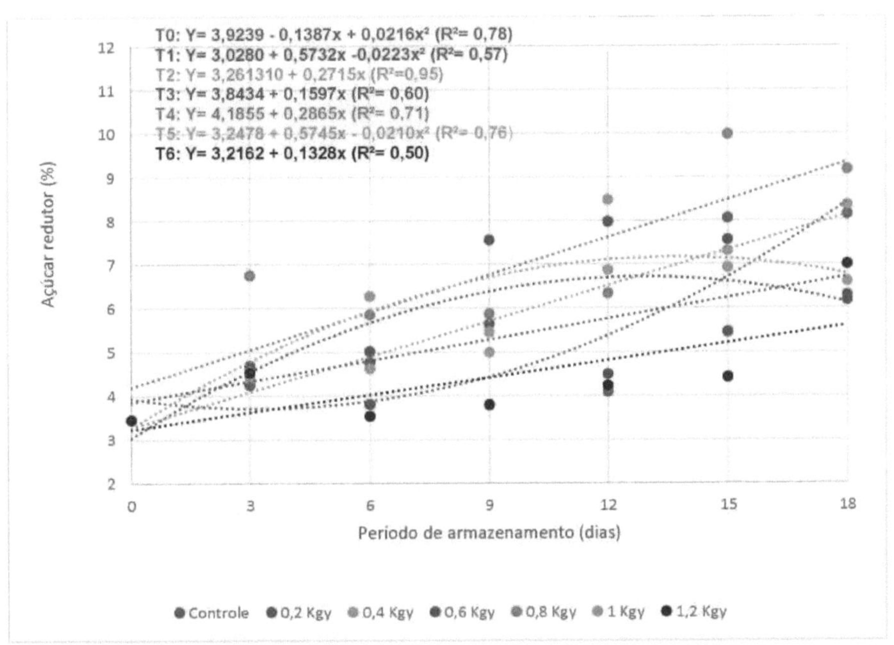

Total sugar in irradiated atemoia

The total sugar content made it possible to assess that the 0.4kGy treatment showed a characteristic increase in its values up to the 12th day of analysis, with 19.22% total sugar on this day, as can be seen in (Figure 29); The evaluation of sugars is important for the quality of the fruit, as they can, for example, contribute to enzymatic actions which can alter the color of the peel and/or pulp in such a way as to negatively influence its appearance, according to the consumer's quality criteria, According to Bruice (2014) total sugars are characterized as monosaccharides, also known as reducing sugars (RA), and the other sugars, such as disaccharides and oligosaccharides, are known as non-reducing sugars (NRA); these sugars interfere with the biochemical process of fruit ripening.

Figura 29 - Total sugar content (%) in atemoya cv. 'Thompson' as a function of irradiation doses (kGy) and packed in expanded polystyrene and polyvinyl chloride trays and stored at 15 ± 0.2°C and 90 ± 2% RH for 18 days. Botucatu SP

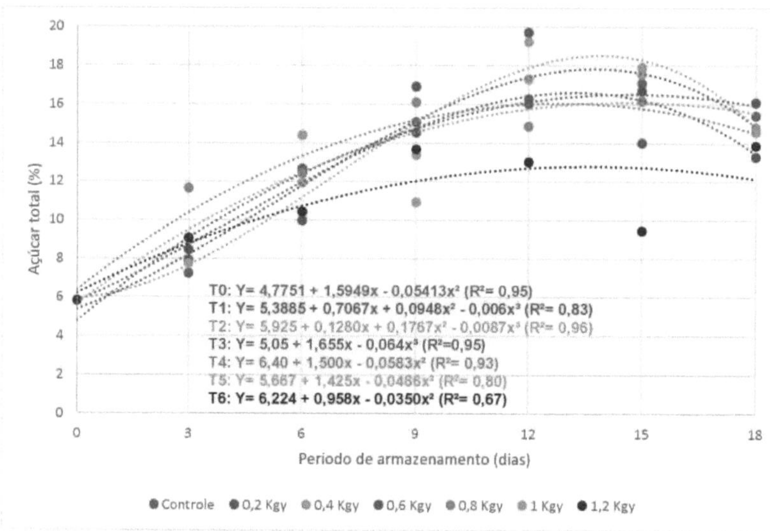

Sucrose in irradiated atemoia

The sucrose content of the atemoyas showed that on the 9th day, the sucrose content found in the 1,OkGy treatment was the lowest among the treatments, according to (Figure 30); This data makes it possible to interpret that this treatment proves to be more efficient in increasing the shelf life of the atemoia, since sucrose is a polysaccharide that can undergo hydrolysis and this biochemical reaction is a factor that favors the ripening of the fruit, contrary data were obtained by Lundgren (2017) who studied the behavior of atemoia cv. 'Thompson' treated with different concentrations of 1-MCP, found a peak in sucrose on the 6th day of analysis, after which the levels remained similar on the 9th and 12th days, decreasing towards the end of the experiment.

Figura 30 - Sucrose content (%) in atemoya cv. 'Thompson' as a function of irradiation doses (kGy) and packed in expanded polystyrene and polyvinyl chloride trays and stored at 15 ± 0.2°C and 90 ± 2% RH for 18

days. Botucatu SP

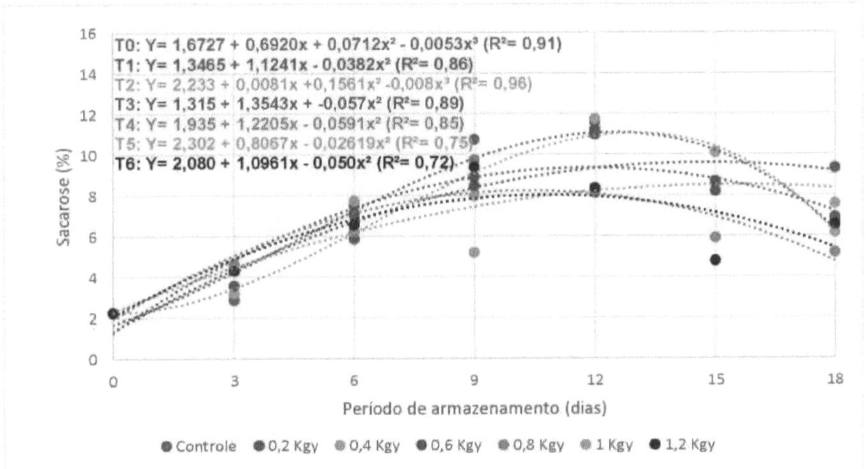

Starch in irradiated atemoya

The starch content of the atemoyas showed a relationship between treatment doses and storage time, characterized by a decrease in starch values over time for all treatments, from day zero to day 18. It was possible to observe that the fruit from the 1.2kGy treatment showed a different behavior, with the lowest average loss during this period; It was also observed that between the 9th and 15th day, it was the treatment with the lowest starch loss of all the other treatments, reaching 4.58% starch on the 15th day of storage, which was 352% higher than the control treatment over the same period, as can be seen in Figure 31. This information makes it possible to interpret that this treatment increased the shelf life of the atemoia, because with the lower loss of starch, the fruit experienced less respiration, reducing the transformation of starch into soluble sugars such as glucose, fructose and sucrose, which increase with respiration, causing the fruit's shelf life to decrease, This statement corroborates Ventura (2010) who mentions that sugar can be hydrolyzed or oxidized when subjected to gamma radiation, which increases the initial level of soluble water, reducing sugar, Calore (2003) studying the application of irradiation to peaches cv. 'Buiuti' at doses of 0.1, 0.2, 0.3, 0.4 and 0.5 kGy found that the different doses of irradiation did not interfere in the process of delaying the ripening of the fruit; Daiuto (2010) observed that the doses of 0.2, 0.4, 0.6 and 1.0 kGy of gamma irradiation on avocado fruit cv. 'Hass' were not efficient in preserving the fruit.

Figura 31 - Starch content (%) in atemoya cv. 'Thompson' as a function of irradiation doses (kGy) and packed in expanded polystyrene and polyvinyl chloride trays and stored at 15 ± 0.2°C and 90 ± 2% RH for 18 days. Botucatu SP

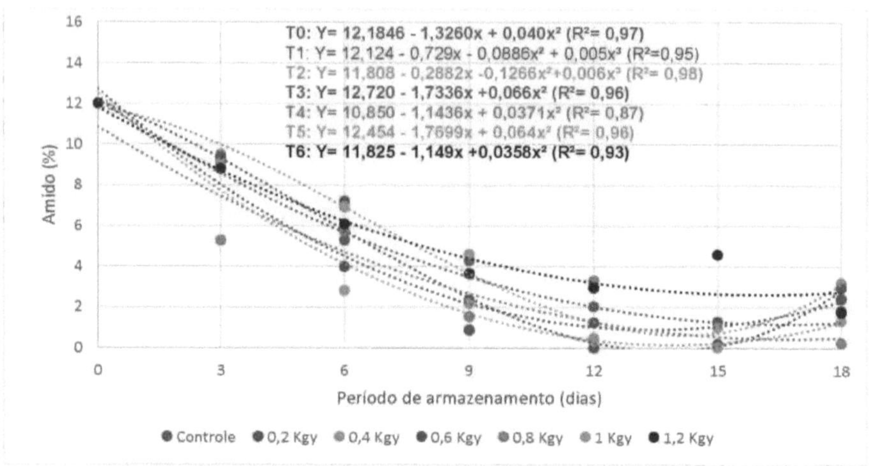

Bark luminosity in irradiated atemoia

The results obtained with the use of ionizing radiation on atemoyas for the shell brightness criterion showed that it was not significant in the treatments, as can be seen in (Table 19).

Table 19 - Luminosity values (L*) of the peel in 'Thompson' cv. atemoia as a function of irradiation doses (kGy) and packed in expanded polystyrene and polyvinyl chloride trays and stored at 15 ± 0.2°C and 90 ± 2% RH for 18 days. Botucatu SP

Doses	Storage period (days)							Average
	0	3	6	9	12	15	18	
Control*	55,73	47,88	52,44	53,67	47,41	47,04	46,89	50,15
0.2 kGy	55,73	47,67	50,51	49,46	48,49	54,61	54,49	51,57
0.4 kGy	55,73	49,35	51,88	48,62	54,76	50,76	45,31	50,91
0.6 kGy	55,73	54,99	48,82	52,59	50,50	46,81	48,34	51,11
0.8 kGy	55,73	47,94	54,64	57,61	51,21	49,22	42,94	51,33
1.0 kGy	55,73	50,96	51,49	53,60	52,71	53,39	46,98	52,12
1.2 kGy	55,73	47,07	49,68	53,95	48,00	51,43	44,97	50,12
C.V. (%)	8,6							

Tukey test.

(*) The control treatment is characterized by the fruit being washed under running water and then dried at room temperature.

The data obtained on the luminosity of the peel of the 'Thompson' cv. "Thompson", related to the averages of the treatments and the storage period, showed in Figure 32, that the variation in the behaviour of luminosity over the storage period was characterized by a decrease in luminosity until the 6th day, with a subsequent increase until the 12th day, a point which determined a further drop in luminosity until the 18th day; This statement can be interpreted as possibly being related to the action of enzymes linked to the fruit's ripening process, which acted during storage; in addition to this data, there is also that observed by Lundgren (2017), who studied the behavior of atemoias cv.'Thompson' treated with 1-methylcyclopropene showed similar behavior for the periods mentioned in this work.

Figure 32 - Luminosity values (L*) of the peel in 'Thompson' cv. atemoia as a function of irradiation doses (kGy) and packed in expanded polystyrene and polyvinyl chloride trays and stored at 15 ± 0.2°C and 90 ± 2% RH for 18 days. Botucatu SP

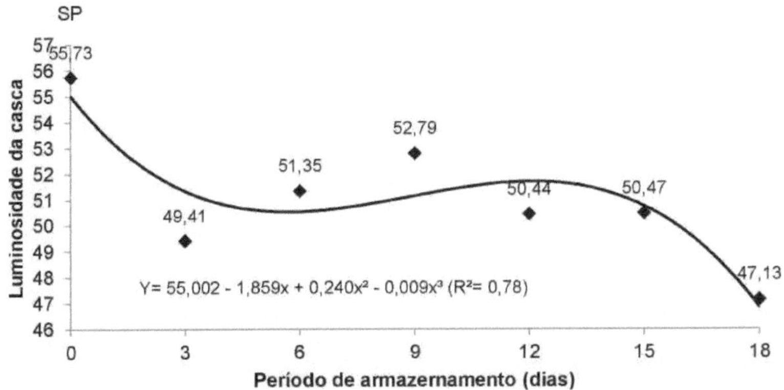

Peel chroma in irradiated atemoia

The data shown in Figure 33 shows that until the 6th day there was a drop in the chroma values of the peel for all the treatments. After this period there were differences between the treatments, with the 1.0kGy treatment showing an increase in the chroma values of the peel on the 15th day, This statement makes it possible to interpret the positive action of ionizing irradiation in delaying the biochemical actions that interfere with the visual quality of the fruit, through enzymatic browning possibly linked to the enzyme peroxidase; This data is in disagreement with Lundgren (2017) who observed that in 'Thompson' cv. atemoias treated with 1-methylester, the color of the fruit is not the same as the color of the fruit. 'Thompson' treated with 1-methylcyclopropene, there was a constant decrease in the chroma values of the peel after the 6th day of storage for all the treatments proposed in the experiment. In an experiment on the use of ionizing irradiation on blueberries, Ventura et al (2010) found that it did not cause a loss of color; these data corroborate the value of the study of colorimetry in this experiment.

Figura 33 - Peel chroma values of 'Thompson' cv. atemoia as a function of irradiation doses (kGy) and packed in expanded polystyrene and polyvinyl chloride trays and stored at 15 ± 0.2°C and 90 ± 2% RH for 18 days. Botucatu SP

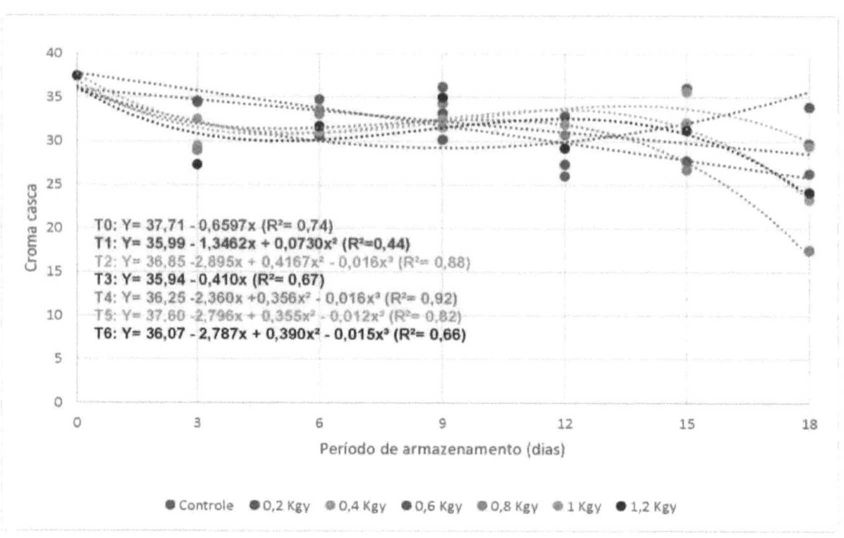

°Hue in irradiated atemoya peel

The ° *Hue* characterization ranges from 0 to 360°, indicating red (0°), yellow (90°), blue (270°) and green (360°) coloration, and the observation of the data in (Table 21), made it possible to interpret the peel tone as indicating a yellowish color on day zero, maintaining this color during storage, which makes it possible to interpret the non-significance of the treatments in the variation of *°Hue* in irradiated atemoyas.

Table 20 - Peel *°Hue* values as a function of irradiation doses (kGy) in 'Thompson' cv. atemoia packed in expanded polystyrene polyvinyl chloride trays and stored at 15 ± 0.2°C and 90 ± 2% RH for 18 days. Botucatu SP

Doses	Storage period (days)							Average
	0	3	6	9	12	15	18	
Control*	97,48	105,9	107,4	103,5	90,8	104,1	102,9	101,7
0.2kGy	97,48	108,4	106,4	108,5	103,7	106,2	111,3	106,0
0.4kGy	97,48	103,6	106,0	110,7	109,6	105,9	92,60	103,7
0.6kGy	97,48	112,2	104,2	109,8	106,2	102,1	101,8	104,8
0.8kGy	97,48	104,2	111,9	108,1	106,1	109,0	90,51	103,9
1.0kGy	97,48	109,1	108,7	110,4	106,7	104,5	102,3	105,6
1.2kGy	97,48	101,9	104,4	108,2	105,9	111,3	100,7	104,3
C.V. (%)	6,8							

The means of treatment and interaction were not significant at 5% probability, according to Tukey's test.

(*) The control treatment is characterized by the fruit being washed under running water and then dried at room temperature.

The data contained in Figure 34 shows the characterization of the averages obtained over the course of the storage period, in terms of the variation in *°Hue*, with the home colour remaining yellowish from day zero to day 9, indicating an increase in the shade of yellow, which then decreased from this point until day 18, This statement makes it possible to interpret that ionizing radiation may have had a positive influence on the color of the skin, possibly by not increasing ethylene, as it interacts with the breakdown of chlorophyll molecules present in the skin of the fruit. Lundgren (2017) obtained similar results in 'Thompsom' atemoyas treated with 1-methylcyclopropene, which showed a decrease in their° Hue values over the 18-day storage period.

Figura 34 - Values of the ° *Hue* of the peel in 'Thompson' cv. atemoia as a function of irradiation doses (kGy) and packed in expanded polystyrene and polyvinyl chloride trays and stored at 15 ± 0.2°C and 90 ± 2% RH for 18 days. Botucatu SP

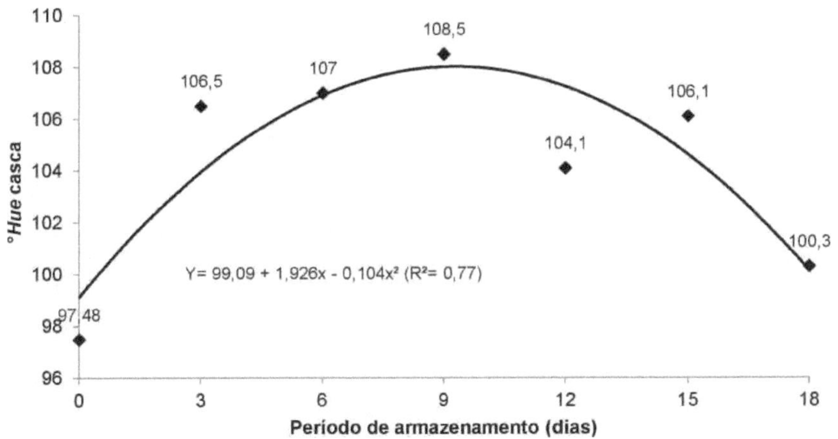

Pulp brightness in irradiated atemoia

The luminosity values of the irradiated atemoya pulp were not significant for the treatments or for the interaction between the treatments and the storage period (Table 22).

Table 21 - Luminosity values (L*) of the pulp as a function of irradiation doses (kGy) in 'Thompson' cv. atemoia packed in expanded polystyrene and polyvinyl chloride trays and stored at 15 ± 0.2°C and 90 ± 2% RH for 18 days. Botucatu SP

Doses	Storage period (days)							Average
	0	3	6	9	12	15	18	
Control*	95,0	82,5	93,9	93,0	93,5	89,3	91,6	91,3
0.2kGy	95,0	95,4	93,6	92,4	85,3	92,0	92,4	92,3
0.4kGy	95,0	95,6	94,6	94,7	92,8	91,1	85,9	92,8
0.6kGy	95,0	95,4	93,6	91,7	92,0	88,9	88,9	92,2
0.8kGy	95,0	94,4	94,9	93,5	93,6	88,4	85,3	92,2
1.0kGy	95,0	95,6	92,4	93,9	90,5	90,7	91,2	92,8
1.2kGy	95,0	94,7	94,3	95,0	94,8	91,6	89,2	93,5
C.V. (%)	4,0							

by the Tukey test.

(*) The control treatment is characterized by the fruit being washed under running water and then dried at room temperature.

The results obtained for the luminosity of the flesh, based on the averages obtained in the treatments, make it possible to interpret these data as the luminosity of the white flesh, taking into account the CIELAB (CIE - Comission Internationale de L'Eclairage) space; a fact that can possibly be explained by the increase in the concentration of the enzyme peroxidase in the flesh of irradiated atemoia fruit during the ripening process, a fact that negatively interferes with visual analysis by the consumer, in terms of quality criteria for consumption.

Figure 35 - Luminosity values (L*) of the pulp in 'Thompson' cv. atemoia as a function of irradiation doses (kGy) and packed in expanded polystyrene and polyvinyl chloride trays and stored at 15 ± 0.2°C and 90 ± 2% RH for 18 days. Botucatu SP

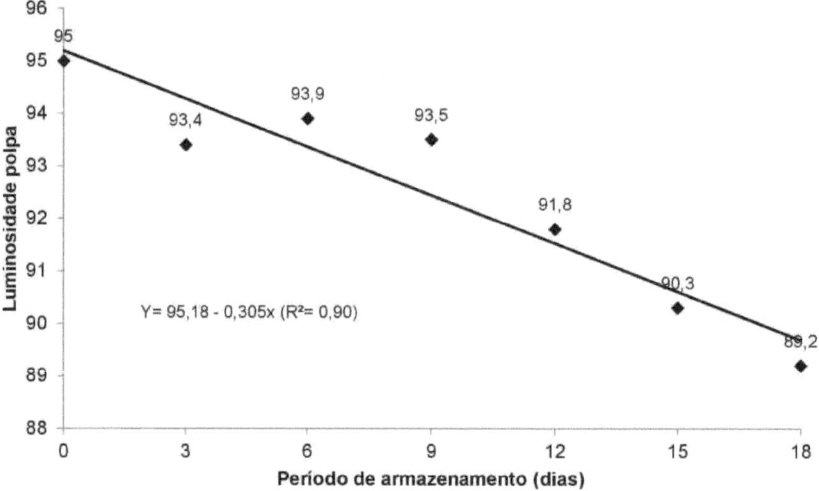

Pulp chroma in irradiated atemoia

The chroma content of the atemoya pulp showed that its saturation was not altered by the doses of irradiation to which the fruit was subjected, information characterized by the values described in (Table 22), this information is confirmed in the analysis of chromatology Minolta (1998), characterizes chroma as a reference value to the distance from the central axis of the CIELAB space, which in this case is characterized by cream-colored chroma.

Table 22 - Pulp chroma values as a function of irradiation doses (kGy) in 'Thompson' cv. atemoia packed in expanded polystyrene and polyvinyl chloride trays and stored at 15 ± 0.2°C and 90 ± 2% RH for 18 days. Botucatu SP

Doses	Storage period (days)							Average
	0	3	6	9	12	15	18	
Control*	11,3	10,8	12,4	13,7	13,7	15,5	15,0	13,2
0.2kGy	11,3	11,5	13,1	13,5	15,9	14,3	14,9	13,5

0.4kGy	11,3	12,8	12,7	13,7	13,9	14,8	15,4	13,5
0.6kGy	11,3	11,3	12,5	15,7	14,4	14,1	17,7	13,9
0.8kGy	11,3	12,6	12,4	14,6	13,1	15,4	17,9	13,9
1.0kGy	11,3	11,4	13,8	13,4	15,6	14,4	15,1	13,6
1.2kGy	11,3	11,8	13,5	12,9	13,8	15,3	14,5	13,3
C.V. (%)	9,7							

The means of the treatment and interaction were not significant at the 5% probability level.

Tukey test.

(*) The control treatment is characterized by the fruit being washed under running water and then dried at room temperature. .

The results shown for the pulp chroma values in atemoyas were characterized by a variation in pulp chroma, in terms of the averages obtained by the treatments during the storage periods. These results showed increasing changes in pulp chroma, exemplified by a 139% differential in the variation between pulp chroma values between day zero and day 18 of the experiment, which resulted in the colour varying from cream chroma to yellow over the course of the 18 days; This statement makes it possible to interpret the non-interaction between treatment and storage period, but storage time as a significant factor influencing the change in pulp chroma, since fruit with high starch and/or sugar content can favor enzymatic activity, such as peroxide, which can favor the change in pulp chroma during the storage period;Chitarra (2005) who explain that the visual quality of fruit post-harvest is an important factor when it comes to the criteria used by consumers to determine the quality of a product.

Figure 36 - Pulp chroma values as a function of irradiation doses (kGy) in 'Thompson' cv. atemoia and packed in expanded polystyrene and polyvinyl chloride trays and stored at 15 ± 0.2°C and 90 ± 2% RH for 18 days. Botucatu SP

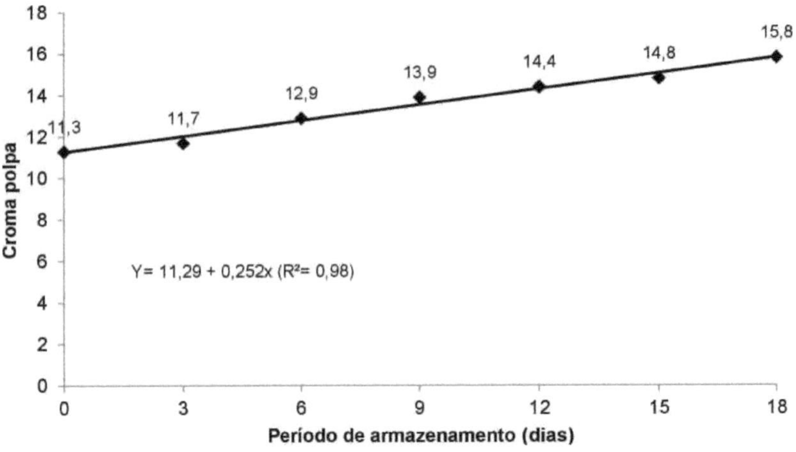

°*Hue* of the pulp in irradiated atemoia

The °*Hue* values of the irradiated atemoya pulp characterized the non-significance of the averages obtained in the treatments as well as in their interaction with storage time, according to the data described in Table 24, which made it possible to interpret that ionizing radiation did not interfere in the biochemical process of the internal colouring of the pulp, for the purposes of characterizing the ripening point, since °*Hue* represents the angle of tone, which starts on the +a* axis, axis and moves counterclockwise and is expressed in degrees (for example, 0° is red and 90° is yellow). This can also be seen in Figure 37, which shows that the averages obtained between treatments and storage periods did not differ statistically from the control treatment, This data corroborates Mendonça (2016), who observed no change in the° Hue for persimmon pulp cv. 'Kioto' when using ionizing radiation as physical methods *in* the preservation of 'Kioto' persimmon cv. *in natura* and or minimally processed; this information makes it possible to interpret that the study of *oHue* contributes to the interpretation of the degree of ripeness of the fruit.

Table 23 - Pulp ° *Hue* values as a function of irradiation doses (kGy) in 'Thompson' cv. atemoia packed in expanded polystyrene and polyvinyl chloride trays and stored at 15 ± 0.2°C and 90 ± 2% RH for 18 days. Botucatu SP

Doses	0	Storage period (days)					18	Average
		3	6	9	12	15		
Control*	91,9	94,8	95,3	95,5	94,5	94,3	94,2	94,3
0.2kGy	91,9	94,2	95,1	96,0	92,1	95,8	96,0	94,5
0.4kGy	91,9	96,1	94,1	95,4	95,2	95,7	89,5	94,0
0.6kGy	91,9	95,6	94,3	94,6	94,1	93,8	88,3	93,2
0.8kGy	91,9	93,0	95,0	95,8	94,6	95,0	92,7	94,0
1.0kGy	91,9	95,5	92,1	94,6	93,7	95,1	90,3	93,3
1.2kGy	91,9	95,0	92,7	94,4	94,1	88,4	92,7	92,8
C.V. (%)	2,7							

The means of treatment and interaction were not significant at 5% probability, according to Tukey's test.

(*) The control treatment is characterized by the fruit being washed under running water and then dried at room temperature.

Figure 37 - °*Hue* values of the pulp in 'Thompson' cv. atemoia as a function of irradiation doses (kGy) and packed in expanded polystyrene and polyvinyl chloride trays and stored at 15 ± 0.2°C and 90 ± 2% RH for 18 days. Botucatu SP

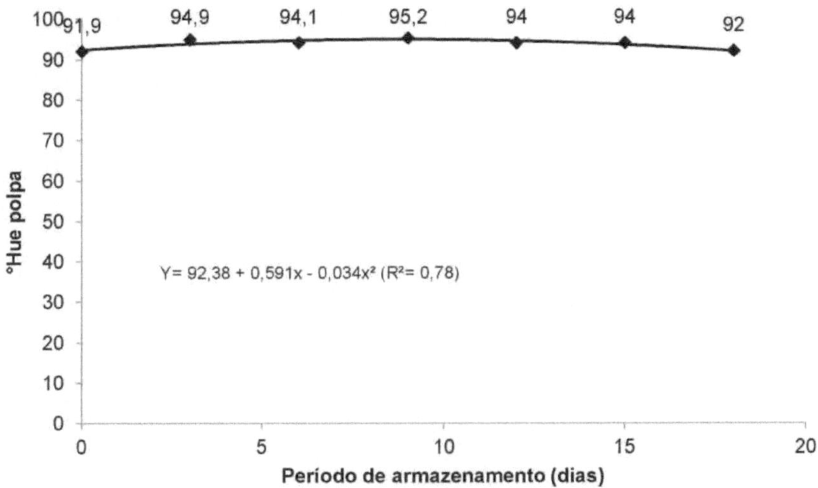

Total antioxidant activity in irradiated atemoia

The peak of total antioxidant activity occurred in the irradiated atemoia on day 6, for the treatment with 1.0kGy reaching 90.9%, in this same period the treatment with 1.2kGy showed 88.84% of total antioxidant activity, later this same treatment maintained a high rate of antioxidant activity until day 9 showing a rate of 93.4% on this day; After this period, the antioxidant activity decreased again and on the 18th day there was a new peak of activity with 92.7%. These data allow us to interpret that the treatment with 1.2kGy favored the atemoyas in delaying the autoxidation stages, by different mechanisms, which may interfere in increasing shelf life; This statement corroborates Castro (2012), who found a correlation between phenol content and antioxidant activity, and this statement corroborates data obtained by Mendonça (2016), who found significant antioxidant activity for the 1.2kGy dose in persimmon cv.'Kioto'.

Figure 38 - Total antioxidant activity (%) in atemoia cv. 'Thompson' as a function of irradiation doses (kGy) and packed in expanded polystyrene and polyvinyl chloride trays and stored at 15 ± 0.2°C and 90 ± 2% RH for 18 days. Botucatu SP

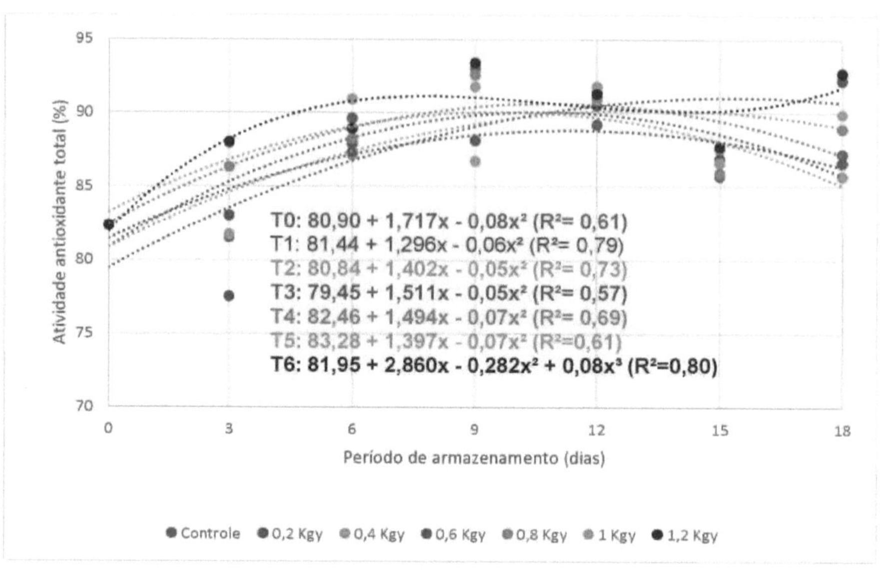

Total phenolic compounds in irradiated atemoia

The results obtained for total phenolic compounds in (Table 24) show that there was no significance between the treatments or in their interaction with the storage period.

Table 24 - Total phenolic compound content (mg gallic acid 100^{-1} g) as a function of irradiation dose (kGy) in 'Thompson' cv. atemoia packed in expanded polystyrene and polyvinyl chloride trays and stored at 15 ± 0.2°C and 90 ± 2% RH for 18 days. Botucatu SP

Doses	Storage period (days)							Average
	0	3	6	9	12	15	18	
Control	2351,8	3642,7	3647,0	2473,7	1525,4	1632,0	2884,3	2593,8
0.2kGy	2351,8	2613,7	4289,4	2194,4	2822,7	1146,2	508,5	2275,2
0.4kGy	2351,8	4943,7	2828,9	2163,3	2803,0	3995,7	403,9	2784,3
0.6kGy	2351,8	2244,5	1300,1	2145,3	2736,6	1395,5	603,9	1825,4
0.8kGy	2351,8	3230,1	1742,8	2948,0	3033,7	2791,7	863,9	2423,1
1.0kGy	2351,8	2100,0	1959,9	1248,0	2502,8	1168,6	1060,8	1770,3
1.2kGy	2351,8	2213,3	1932,9	1768,9	2246,1	2417,8	911,6	1977,5
C.V. (%)	62,1							

The means of treatment and interaction were not significant at 5% probability, according to Tukey's test.

The control treatment consists of fruit washed in running water and then dried at room temperature.

The results shown in Figure 39 allow us to interpret that from the 3rd day of storage onwards, the averages obtained for the total phenolic compound content (mg of gallic acid 100^{-1} g) showed a downward trend; This fact is coupled with a decrease in antioxidant action in the atemoya, which is characterized, for example, by a decrease in ascorbic acid and an increase in tissue oxidation. This statement corroborates Rosenthal (2008), who explains that free radicals can react with other molecules, altering the aroma of the fruit through the production of volatile compounds, This statement confirms Brand-Williams (1995), who characterizes the decrease in the ability to inhibit or delay oxidation by inactivating free radicals thanks to the donation of hydrogen atoms or electrons in cells, transforming free radicals into stable substances; data obtained by Lundgren (2017), showed a better response of phenolic compounds in atemoya cv. 'Thompsom' on the 6th day of analysis, when subjected to treatment with 1-methylcyclopropene (200 µL.L-1, 300 µL.L-1 and 400 µL.L-1) and stored under refrigeration in a cold room at 15°C ± 1 and 90 ± 5% RH, Lima (2012) obtained a reduction in the flavonoid content, in relation to the control sample, for samples of sugarcane juice irradiated with cobalt-60 at doses of 5 and 10kGy, with the same author reporting that the higher the doses applied, the higher the content of phenolic compounds found, with the highest dose applied being 20kGy.

Figure 39 - Total phenolic compound content (mg of gallic acid 100^{-1} g) in 'Thompson' atemoia as a function of irradiation dose (kGy) and packaging in expanded polystyrene and polyvinyl chloride trays and stored at 15 ± 0.2°C and 90 ± 2%.

of RH for 18 days. Botucatu SP

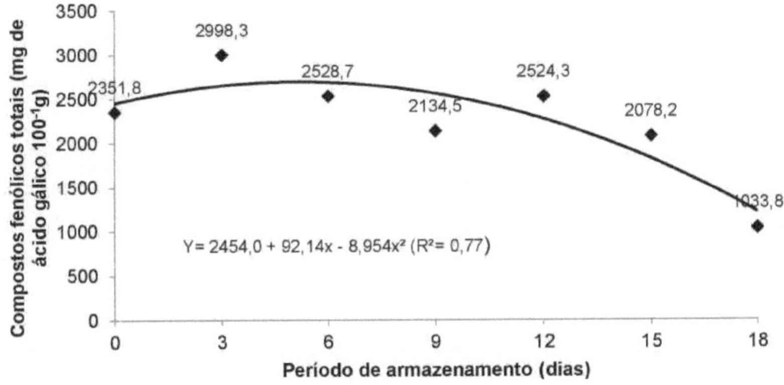

Polyphenoloxidase (PPO) in irradiated atemoia

The data obtained on the influence of ionizing irradiation on 'Thompson' cv. atemoias showed that there was no statistically significant difference between the treatments. "Thompson" for the activity of polyphenol oxidase (PPO) showed in Table 25 that there was no statistically significant difference between the treatments, as well as their interaction with the storage period; This makes it possible to interpret, in this experiment, the lack of influence of gamma ray irradiation (Cobalt 60) as a method of controlling enzymatic processes which negatively affect the shelf life of the fruit, causing greater post-harvest losses.

Table 25 - Values of polyphenol oxidase activity (μmol catechol transformed min^{-1} g^{-1} of fresh matter) as a function of irradiation doses (kGy) in 'Thompson' atemoia cv. packed in expanded polystyrene and polyvinyl chloride trays and stored at 15 ± 0.2°C and 90 ± 2% RH for 18 days. Botucatu SP

Doses	Storage period (days)							Average
	0	3	6	9	12	15	18	
Control*	0,420	0,550	0,227	0,390	0,270	0,213	0,120	0,313
0.2kGy	0,420	0,660	0,243	0,220	0,447	0,430	0,247	0,381
0.4kGy	0,420	0,410	0,390	0,280	0,380	0,173	0,270	0,332
0.6kGy	0,420	0,520	0,510	0,160	0,330	0,187	0,157	0,326
0.8kGy	0,420	0,327	0,610	0,253	0,343	0,280	0,237	0,353
1.0kGy	0,420	0,457	0,567	0,413	0,233	0,323	0,443	0,408
1.2kGy	0,420	0,434	0,450	0,220	0,280	0,123	0,270	0,314
C.V. (%)	49,4							

The means of treatment and interaction were not significant at 5% probability, according to Tukey's test.

(*) The control treatment is characterized by the fruit being washed under running water and then dried at room temperature.

Looking at Figure 40, it is possible to interpret the influence of the storage period in relation to the action of the PPO enzyme; the averages obtained over the days of analysis show that from the 6th day onwards there was a reduction in the activity of the PPO enzyme, which makes it possible to interpret the action of bioactives in controlling it; This observation can be complemented by Chitarra and Chitarra (2005) who cite the use of doses above 1.0kGy as being higher than the dose tolerated by the plant tissue, allowing it to suffer a physiological disorder; This could alter the permeability of the cell membrane, resulting in greater entry of molecular oxygen into the cytoplasm and thus causing a decompartmentalization of phenolic compounds, causing their oxidation; This statement corroborates Beaulieu, D'Aprano and Lacroix (2002) who state that gamma radiation can directly affect enzymatic browning by altering the protein structural conformation of enzymes or modifying their active center, Silva (2013), studying the preservation of juçara pulp using gamma radiation at doses of 2.5kGy and 5.0kGy, observed that there was no variation in the activity of the PPO enzyme over the analysis periods of 1, 4, 21 and 28 days.

Figure 40 - Polyphenol oxidase activity (μmol catechol transformed min^{-1} g^{-1} of fresh matter) in 'Thompson' atemoia as a function of irradiation doses (kGy) and packed in expanded polystyrene and polyvinyl chloride trays and stored at 15 ± 0.2°C and 90 ± 2% RH for 18 days. Botucatu SP

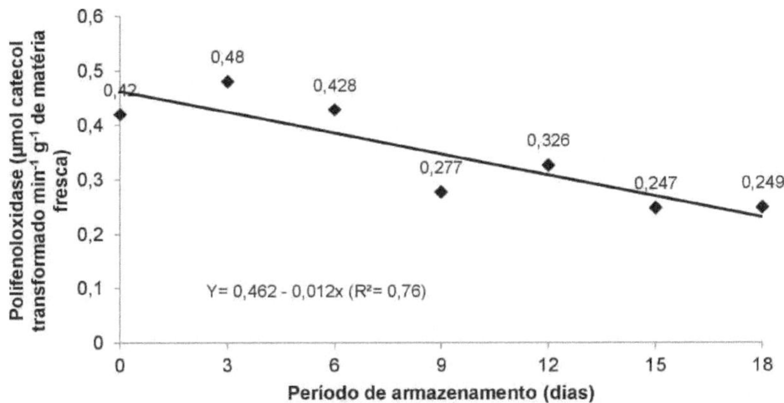

Peroxidase (POD) in irradiated atemoia

The peroxidase activity in atemoya fruit was reduced with the doses of 0.2kGy and 1.2kGy, which represented values 10% and 28% lower than the control treatment, respectively, as shown in Figure 41, These statements make it possible to interpret that the doses of 0.2kGy and 1.2kGy favored a better behavior of the atemoya in terms of maintaining the color of the fruit compared to the other treatments, Chitarra and Chitarra (2005) report that ionizing radiation can generate a reaction with the water in the tissues and molecular oxygen, which reacts with the hydrated electron to form superoxide, which can then form hydrogen peroxide ($2O_2^- + 2H^+ \rightarrow H2O2 + O2$), This statement corroborates the data obtained by Fujita, Vieites and Campos (2012) who found no influence of gamma irradiation doses (0.0, 0.2, 0.4, 0.6 and 0.8 kGy) on the reduction of peroxide activity in manna cubiu fruit; These statements make it possible to interpret the importance of this experiment in studying different methodologies aimed at reducing post-harvest losses.

Figure 41 - Peroxidase activity (µmol H2O2 transformed min^{-1} g^{-1} fresh matter) in 'Thompson' atemoia as a function of irradiation doses (kGy) and packed in expanded polystyrene and polyvinyl chloride trays and stored at 15 ± 0.2°C and 90 ± 2% RH for 18 days. Botucatu SP

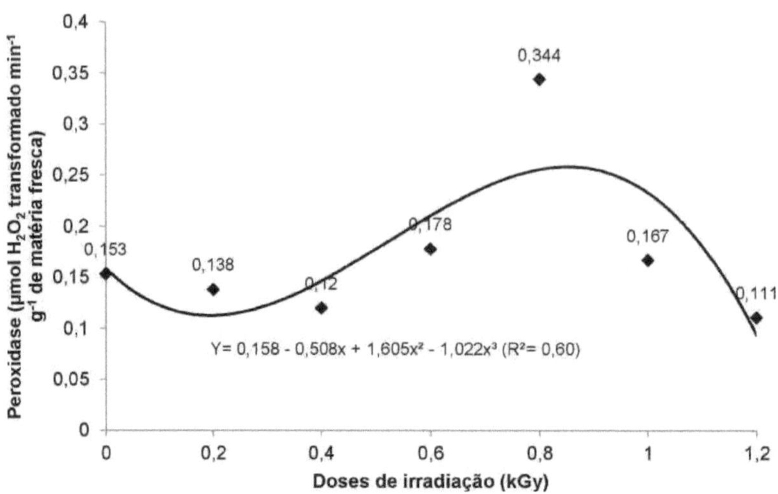

The peroxidase activity during the storage period of the irradiated atemoya fruits showed a variation in POD over the 18 (eighteen) days of storage, showing a difference of 377% more between day zero and day 18, according to (Figure 42). These statements make it possible to interpret that there was no interaction between the doses of ionizing irradiation and the storage period in terms of peroxidase inactivation; This statement makes it possible to interpret that ionizing irradiation did not favor the production of bioactives that act in the inactivation of free radicals, responsible for accelerating cell senescence; which corroborates Lundgren (2017) who found that the enzymatic activity of peroxidase in atemoya cv. 'Thompson', treated with different doses of 1-MCP, evolved continuously during the storage period, with its peak concentration occurring on the 9th day. These data differ from Torres (2008), who observed an increase in peroxidase activity in atemoya fruit cv. "Thompson" until the 6th day of analysis, when they were immersed in calcium chloride at temperatures of 20° and 40° for 20 minutes, characterizing the darkening of the pulp by the action of this enzyme; these statements corroborate the importance of this study in the search for solutions to post-harvest losses.

Figure 42 - Peroxidase activity (μmol H2O2 transformed min^{-1} g^{-1} fresh matter) in 'Thompson' atemoia as a function of irradiation doses (kGy) and packed in expanded polystyrene and polyvinyl chloride trays and stored at 15 ± 0.2°C and 90 ± 2% RH for 18 days. Botucatu SP

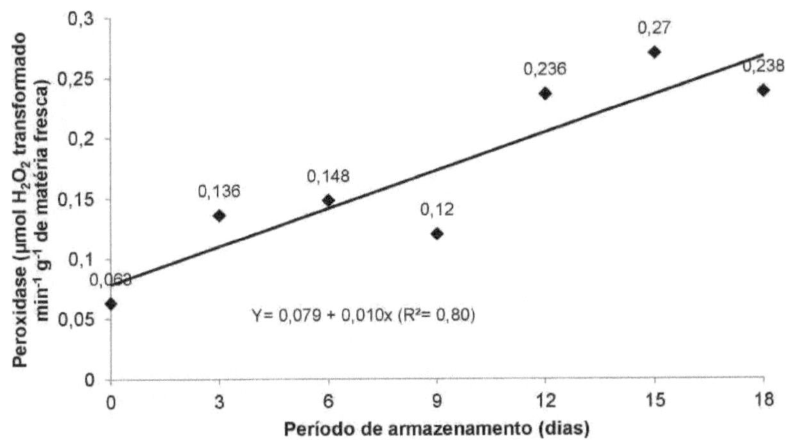

2.4 CONCLUSION

The results obtained in this experiment allow us to conclude that:

- The doses of 1.0kGy and 1.2kGy show greater control of respiration in the atemoya up to the 9th day, three days longer than the control treatment.
- The use of ionizing irradiation did not activate the biosynthesis of bioactives in the atemoia; a fact possibly related to the phytosanitary quality of the orchard, transport boxes and storage, which possibly reduced contaminating agents.

REFERENCES

AOAC. **Official methods of analysis of the Association of Official Analytical Chemistry International**. 13th ed. Washington, 1015 p., 1992.

AZEVEDO, Fabiana Garcia de. **Gamma Irradiation in Food**. 2009. 29 f. Monograph (Specialization) - Agronomy Course, Energy in Agriculture, Universidade Estadual Paulista "Jùlio de Mesquita Filho" Faculdade de Ciências Agronômicas Campus de Botucatu, Botucatu, 2θ09.

BEAULIEU, M.; D'APRANO, G.; LACROIX, M. Effectofdose Pateofgammairradiation on biochemical quality and browning of mushrooms Agaricus bisporus. **Radiation Physics and Chemistry**, v.63, n.3-6, p.311-315, maarch, 2002.

BLEINROTH, E. W.; ZUCHINI, A. G.; POMPEO, R. M. Determination of the physical and mechanical characteristics of avocado varieties and their cold storage. **Coletânea ITAL,** Campinas. v. 7, n. 1, p. 29-81, 1976.

BRAND-WILLIAMS, W.; CUVELIER, M. E.; BERSET, C. Use of a free radical method to evaluate antioxidant activity. Lebensmittel-Wissenschaft Technologie, London, v. 28, p. 25-30, 1995.

BRAZIL. Adolfo Lutz Institute (IAL). Normas Analiticas do Instituto Adolfo Lutz - **Métodos Quimicos e Fisicos para Analise de Alimentos**. 4ª edition, Brasilia: 2008.1018p.

BRUICE, Paula Y. **Fundamentals of organic chemistry**. 2. Ed. São Paulo: Pearson Education do Brasil, 2014.

CALORE, Luciana; VIEITES, Rogério Lopes. Preservation of 'Biuti' peaches by irradiation. **Ciência Tecnologia de Alimentos,** v.2, n. 23, p. 53-57, dec. 2003.

CANO, M. P. *et al* Differences among Spanish and Latin-American banana cultivars: morphological, chemical and sensory characteristics. **Food Chemistry.** n. 59, p. 41119, Amsterdam,1997.

CARDELLO, Maria Helena A. B.; CARDELLO, Leonardo. VITAMIN C CONTENT, ASCORBATE OXIDASE ACTIVITY AND SENSORY PROFILE OF MANGOES (*Mangifera indica* L.) VAR. HADEN, DURING RIPENING. **Ciência e Tecnologia de Alimentos,** Campinas, v. 18, n. 2, p. 213-227, July, 1998.

CARVALHO CRL; MANTOVANI DMB; CARVALHO PRN; MORAES RMM. **Chemical analysis of foods:** technical manual. Campinas: Institute of Food Technology. 121p., 1990.

CASTRO, Juliana Fernanda Almeida. **Study of antioxidant activity in native and exotic Brazilian fruits.** 2012. 86 f. Dissertation (Master's Degree) - Chemistry Course, Department of Biochemistry and Chemical Technology, Universidade Estadual Paulista "Jûlio de Mesquita Filho" Instituto de Quimica, Araraquara, 2012.

CHITARRA M. I. F.; CHITARRA A. B. **Post-harvest of fruits and vegetables - Physiology and Handling.** Lavras: UFLA. 785p, 2005.

COSTA, S. M. **Refrigerated preservation of 'Tropic Beauty' peaches irradiated with and without the use of potassium permanganate.** 2008, 60p. Dissertation (Master's Degree in Agronomy / Horticulture) Faculty of Agronomic Sciences - UNESP, Botucatu, 2008.

DAIUTO, Èrica Regina et al. RESPIRATORY RATE OF 'HASS' ABACADO SUBMITTED TO DIFFERENT PHYSICAL TREATMENTS. **Revista Iberoamericana de Tecnologia** Postcosecha, México, v. 10, n. 2, p.101-109, 2010.

EVANGELISTA, José. **Tecnologia de Alimentos.** 2. ed. São Paulo: Atheneu, 2008. 652 p.

FELLOWS, P.J. **Food Processing Technology**: Principles and Practice. 2. ed. Porto

Alegre: Artmed, 2006. 602 p.

FERREIRA, D. F. **Sisvar** - System for Analysis of Variance. Version 5. 3. Lavras MG: UFLA, 2010.

FONSECA, Paulinie Adenice Quintiliano. **Physico-Chemical Analysis of Fruit Pulp and Evaluation of its Identity and Quality Standards.** 2012. 62 f. Dissertation (Master's Degree) - Chemistry Course, Federal University of Rio Grande do Norte, Natal, 2012.

FUJITA, Erika; VIEITES, Rogério Lopes; CAMPOS, André José de. Post-harvest enzymatic evaluation of cubiu mango treated with gamma radiation. **Revista Agro@mbiente On-line**, Roraima, v. 6, n. 1, p. 53-58, abr. 2012.

GAVA, Altanir Jaime; SILVA, Carlos Alberto Bento da; FRIAS, Jenifer Ribeiro Gava. **Food Technology:** Principles and Applications. Sâo Paulo: Nobel, 2009. 511 p.

KOBLITZ, Maria Gabriela Bello. **Food Biochemistry.** Rio de Janeiro: Guanabara Koogan Ltda, 2008. 242 p.

LEVINE, M.; PADAYATTY, S.J.; ESPEY, M.G. Vitamin C: A Concentration-Function Approach Yields Pharmacology and Therapeutic Discoveries - a review-. **American Society for Nutrition**, v. 2, n. 2, p. 78-88, 2011.

LIMA, G. P. P.; BRASIL, O. G.; OLIVEIRA, A. M. de. Polyamines and peroxidase activity in bean (Phaseolus vulgaris L.) grown under saline stress. **Sci. Agricola**, v. 56, n. 1, 1999.

LIMA, Roberta Bergamin. **Clarification process for sugar cane juice using accelerated electrons.** 2012. 62 f. Dissertation (Master's Degree) - Nuclear Technology Course, Nuclear and Energy Research Institute, University of Sao Paulo, Sao Paulo, 2012.

LUNDGREN, Giovanna Alencar. **CONSERVATION OF ATEMOIA SUBJECTED TO 1-METHYLCYCLOPROPENE.** 2017. 79 f. Dissertation (Master's Degree) - Course in Agricultural Engineering, Horticulture, Faculty of Agronomic Sciences, Unesp - Botucatu Campus, Botucatu, 2017.

MENDONÇA, Veridiana Zocoler de. **Physical methods in the preservation of persimmon cv. Kioto in natura and minimally processed.** 2016. 125 f. Tese (Doutorado) - Agronomy Course, Energy in Agriculture, Universidade Estadual Paulista "Jùlio de Mesquita Filho" Faculdade de Ciências Agronômicas Campus de Botucatu, Botucatu, 2016.

MENSOR, L. L.; MENESES, F. S; LEITÂO, G. G.; REIS, A. S.; SANTOS, T. C.; COUBE, C. S.; LEITÂO, S. G Screening of Brazilian Plant Extracts for Antioxidant Activity by the Use of DPPH Free Radical Method, **phytotherapy Research**, v. 15, n. 2, p. 127-130, 2001.

MINOLTA, K. **Precise color communication: quality control from perception to instrumentation**. 1998. 59p.

NELSON, N. A. A photometric adaptation of somogyi method for determination of glucose. **Journal Biological chemistry**, Baltimore, v.135, n. 5, p. 375, 1944.

NEVES, L.C.; MANZIONE, R.L.; VIEITES, R. L. GAMA RADIATION IN THE POST-HARVEST CONSERVATION OF FRIGOCONSERVED NECTARINE (*Prunus persica* variety 'Nucipersica'). **Revista Brasileira de Fruticultura**, Jaboticabal, v. 24, n. 3, p. 676-679, December 2009.

O'BEIRNE, D. Irradiation of fruits and vegetables: applications and issues. **professional Horticulture**, Oxford, v. 3, n. 1, p. 12-19, 1989.

OH, S. H.; LEE, Y. S.; KIM, J. H.; KIM, J. H.; LEE, J. W.; KIM, M. R.; YOOK, H, S.; BYUN, M. W. Effects of pH on non-enzymatic browning reaction during y-irradiation processing using sugar and sugar-glycine solution. **Food chemistry**, v.94, p-420-427, Amsterdan, 2006.

PEREDA, Juan A. Ordónez et al. **Food Technology: Food Components and Processes.** Porto Alegre: Artmed, 2007. 294 p. Traduçâo Fàtima Murad - Porto Alegre; Artmed, Reprint 2007.

RIBEIRO, E. P.; SERAVALLI, E. A. G. **Quimica de alimentos**. 2ª ed. Sao Paulo. Editora Edgard Blücher Ltda, 196 pages, 2007.

ROSENTHAL, A. **Food technology and innovation: trends and perspectives**. Brasilia: Embrapa Informaçâo Tecnològica, 2008.

SILVA, Ana Veruska Cruz da; MUNIZ, Evandro Neves. QUALITY OF ATEMOYA HARVESTED AT TWO STAGES OF RIPENESS. **Revista Caatinga**, Mossoró, v. 24, n. 4, p. 9-13, 2011.

SILVA, Paula Porrelli Moreira da. **Preservation of juçara pulp *(Euterpe edulis)* subjected to gamma radiation, pasteurization, freeze-drying and atomization.** 2013. 257 f. Thesis (PhD) - Science Course, Center for Nuclear Energy in Agriculture, University of Sao Paulo, Piracicaba, 2013.

SINGLETON, V.L.; ORTHOFER, R.; LAMUELA, R.M. Analysis of total phenols and other oxidation substrates and antioxidants by means of Folin-Ciocalteau reagent. Methods of Enzymology, New York, v.299, p.152-178, 1999.

SOMOGYI, M. Determination of blood sugar. **Journal Biologic Chemical**, n. 160, p.69-73,

Washington DC, 1945.

SOUSA, F. C.; SOUZA, E. P.; CRUZ, C. S. A; GOMES, J. P.; ALMEIDA, F. A. C. PHYSICAL AND PHYSICAL CHEMICAL PARAMETERS OF ATEMOIA 'GEFNER' AT DIFFERENT STAGES OF MATURATION. **Revista Brasileira de Produtos Agroindustriais**, Campina Grande, v. 15, n. 3, p. 329-324, 2013.

TEZOTTO, Jaqueline Uliana V. et al. Gamma Radiation in Plant Products. **Revista Virtual de Quimica**, Piracicaba, v. 7, n. 1, p. 267-277, nov. 2014.

TORRES, Liz Maria Abi Rached. **POST-HARVEST CONSERVATION OF ATEMOIA cv. 'THOMPSON'.** 2008. 149 f. Dissertation (Master's Degree) - Pharmacy Course, Food Science, Faculty of Pharmaceutical Sciences, Universidade Estadual Paulista "Jùlio de Mesquita Filho", Araraquara1 2008.

VENTURA, Diana et al. **USE OF IRRADIATION IN FOOD** PROCESSING: GENERAL FOOD PROCESSING - MODULE II. Coimbra: Coimbra School of Agriculture, 2010. 31 p. Available at: <http://www.esac.pt/noronha/pga/0910/trabalho_mod2/irradiacao_grupo4_t2_word.p df>. Accessed on: October 31, 2017.

VIEITES, R. L. **Post-harvest preservation of tomatoes using gamma radiation, wax and polyethylene bags, stored in refrigerated and ambient conditions.** 1998. 131f. Thesis. Tese (Livre-Docência) - Faculty of Agronomic Sciences, Universidade Estadual Paulista, Botucatu, 1998.

VIEITES, R. L.; EVANGELISTA, R.M.; CAMPOS, A.J. Effect of packaging and gamma irradiation on the control of microbial contamination of minimally processed mango. **Semina: Ciências Agràrias**, Londrina, v. 25, n. 3, p. 197-206, 20 aug. 2004.

CHAPTER 3

Use of Active Modified Atmosphere in the post-harvest of the 'Thompson' atemoya cultivar

SUMMARY

The experiment in question studied the use of active modified atmosphere in 'Thompson' cv. atemoia with the aim of evaluating its influence on delaying respiration processes. The study methodology was based on four treatments plus the control: T0 - 21% O_2; 0.03% CO_2, 78% N_2, T1 - 4% O_2; 5% CO_2, 91% N_2, T2 -4%O_2; 6% CO_2, 90% N_2, T3 - 4% O_2; 7% CO_2, 89% N_2 and T4 - 4% O_2; 8% CO_2, 88% N_2, three replicates were carried out for each treatment with two fruits per replicate. The fruit from each replication was packed in 14-micron nylon-polyethylene plastic packaging measuring 40 cm x 29.5 cm, except for the control treatment which was packed in expanded polystyrene trays sealed with 0.020 mm polyvinyl chloride film. During the experiment, all treatments were stored in a cold room under conditions of 15 ± 0.2°C and 90 ± 2% RH. The atemoias were analyzed at three-day intervals (3, 6, 9, 12, 15, 18) and all treatments were analyzed for: peel and flesh color, loss of fresh mass, soluble solids (SS), titratable acidity (TA), pH, ascorbic acid (AA), reducing sugars, respiration rate, specific polyphenol oxidase (PPO) and peroxidase (POD) activity, total antioxidant activity and total phenolic compounds. The experiment used a completely randomized design (DIC) in a 5 x 7 factorial scheme. The data was submitted to analysis of variance using the F test at a 95% probability level. Based on the results found, it can be concluded that the lowest respiration peak in the atemoyas occurred on the sixth day with treatment T4 = 4% O_2 and 8% CO_2 (55.178 mL of CO_2 Kg of fruit^{-1} hour^{-1}), the refrigeration conditions adopted in this experiment may have been sufficient for the results obtained.

Keywords: *Annona x atemoya* Mabb., technology, preservation, packaging, quality

3.1 INTRODUCTION

The atemoia is a fruit that thrives in tropical and sub-tropical high-altitude climates. This characteristic is related to its hybridization, which involves an annonaceous plant from a temperate climate, the cherimoia, and another from a tropical climate, the pine cone, also known as ata or fruta do conde, both climacteric fruits.

The Annonaceae family has more than 40 genera. In Brazil, the species of the Annona genus are the most widely cultivated, both for *fresh* and industrial consumption.

The atemoia is a small tree with fruit ranging from 200 to 450 grams; it behaves climatically and varies in color from green to yellowish-green during its ripening process.

The ripening of the atemoya presents a period in which a series of biochemical changes occur, called the climacteric peak, which coincides with physical, biochemical and enzymatic changes that influence the quality for consumption. The time it takes for the fruit to reach this climacteric peak is directly related to the environmental conditions to which the fruit is subjected during its ripening process, such as temperature, humidity, oxygen concentrations and atmospheric carbon dioxide.

For atemoyas, the enzymatic browning that develops during ripening is a problem that needs to be solved in the post-harvest period; it requires studies aimed at positively interfering in events that can guarantee the delay of the process in order to increase the period of distribution and consumption of atemoyas.

Active modified atmosphere technology makes it possible to act on the fruit's respiratory process, with the aim of controlling ripening. Active modified atmosphere technology can use plastic packaging, which originates from polymers, which chemically are high molecular weight chemical groups characterized by being flexible, having a thickness of between 0.254 mm and 300 mm, which makes them thin enough to be flexible and foldable without tearing or breaking (CHITARRA; CHITARRA, 2005), reaffirming this information Yamashita et al. (2002) state that packaging fruit in plastic films can reduce the rates of respiration, transpiration, microbial growth and other metabolic reactions that occur in the product, by creating an optimum micro-atmosphere; one way of obtaining an active modified atmosphere is to insufflate an atmosphere made up of gas mixtures inside the package, in pre-established concentrations, promoting moderate vacuum in the package containing the product, and injecting the desired gas mixture before sealing it; thus making it possible to influence the fruit's ability to breathe, acting to control its ripening process.

The positive effect of this action in reducing respiration is to act on ethylene synthesis, causing delayed ripening.

Ethylene can have a desirable effect on fruit, such as uniform ripening, but its effects can also be undesirable when in high concentration, speeding up fruit ripening too much and anticipating senescence. Reduced O_2 availability and high CO_2 levels reduce the hormone's action (KOBLITZ, 2008).

This delay is achieved with an O_2 concentration of less than 10%; however, a minimum range of 1% to 3% is necessary in order to avoid the transformation of aerobic respiration into anaerobic respiration; together with O_2, a CO_2 concentration (> 10%) can have a similar effect to that caused by anaerobic conditions. 10%) can have an effect similar to that caused by anaerobic conditions, since levels of 5 to 10% CO_2 reduce respiratory activity and delay the onset of ripening in climacteric fruit; levels above 10% CO_2 cause tissue damage and anaerobiosis can produce acetaldehydes, which are harmful to tissues, causing their death (CHITARRA; CHITARRA, 2005; STEFFENS et al., 2009; LANA AND FINGER, 2000; BEAUDRY, 2000; KADER, 1989; SALTVEIT, 2003).

Storing strawberry fruit in a modified atmosphere has been shown to increase the concentration of CO_2, which indirectly reduces the incidence of pathogens; which are able to influence the durability of the fruit in post-harvest, depending on the species of product and the pathogen involved (VIEITES et al., 2006).

The use of nylon-polyethylene packaging to characterize the controlled atmosphere *in fresh* 'Kioto' persimmons proved to be efficient for an atmosphere composed of 7% CO_2 and 4% O_2, which enabled the fruit to be preserved for 35 days (MENDONÇA, 2016).

3.2 MATERIAL AND METHOD

The 'Thompson' atemoyas were subjected to different concentrations of active modified atmosphere, in order to evaluate the influence of this methodology in delaying the respiratory process and its influence on quality criteria during storage.

The study was based on three treatments characterized by pre-established atmospheres in cylinders purchased from White Martins®. In the control treatment, the atemoias were packed in expanded polystyrene trays (2 fruits/tray) and sealed with 0.020mm polyvinyl chloride films, under ambient conditions.

The treatments followed the concentrations of: T_0 (control) - 21% O_2; 0.03% CO_2, 78% N_2, T_1 - 4% O_2; 5% CO_2, 91% N_2, T_2 - 4% O_2; 6% CO_2, 90% N_2, T3 - 4% O_2; 7% CO_2, 89% N2 and T4 - 4% O_2; 8% CO_2, 88% N2.

For each treatment, three replications were assigned and for each repetition, two thistles.

The atemoias from each treatment (except the control treatment) and their replications were packed in 14-micron nylon-polyethylene plastic containers measuring 40 cm x 29.5 cm (Photo 6), and a Teqmaq® model AP 500 sealer was used to inject the gases.

All the laboratory analyses took place in the post-harvest laboratories of the Faculty of Agronomic Sciences - FCA - "Júlio de Mesquita Filho" UNESP - Botucatu Campus, SP, and were stored in a cold room at 15 ± 0.2°C and 90 ± 2% RH during the analysis intervals corresponding to: 0, 3, 6, 9, 12, 15 and 18 days. The analysis methodologies followed those described in items 1.2.1 to 1.11.2 of Chapter 1 (Use of organic acids (ascorbic and citric) in the post-harvest period of the 'Thompson' cultivar of atemoya).

3.2.1 Sorting, transporting and receiving the atemoias

All the procedures complied with the methodology criteria described in item 1.2 of Chapter 1 (Use of organic acids (ascorbic and citric) in the post-harvest period in the 'Thompson' cultivar of atemoya).

Photo 6 - Atemoia cv. 'Thompson', packed in active modified atmosphere.

3.3 RESULTS AND DISCUSSION

The results obtained with the respiration of the atemoia made it possible to observe that the control treatment showed a respiration peak on the 3rd day, while the other treatments showed a respiration peak on the 6th day, with emphasis on the treatment with 4% O_2 and 8% CO_2 (55.178 mL of CO_2 Kg of fruit^{-1} hour^{-1}), which showed the lowest respiration index for the period, according to (FIGURE 43). This indicates lower levels of ethylene, a hormone that acts directly on ripening. This information makes it possible to interpret that the use of active modified atmosphere interfered positively in slowing down the respiratory process during storage of the atemoia. This statement can be clarified by Chitarra and Chitarra (2005) who characterize the increase in CO_2 concentration during storage as influencing the reduction in fruit respiration, which slows down the ripening process, This statement is supported by Reis (2013) who obtained the best results in the preservation of 'Gefner' cultivar atemoias stored at 15°C and 85 ± 5% RH under modified atmosphere conditions in active packaging containing an ethylene absorber (potassium permanganate $KMnO_4$), Mosca (2003) studying the respiratory behavior of atemoias cv. 'Gefner', at room temperature, obtained the highest respiration activity on the 7th day of storage, 25.68 mg $CO2._{kg^{\wedge 1}}\,._{h^{\wedge 1}}$, which differs from this work, possibly due to the difference in the methodology studied.

Figure 43 - Respiratory activity (mL of CO_2 Kg of fruit^{-1} hour^{-1}) in atemoia cv.' Thompson' characterized by different gas concentrations in active modified atmosphere and stored at 15 ± 0.2°C and 90 ± 2% RH for 18 days. Botucatu SP

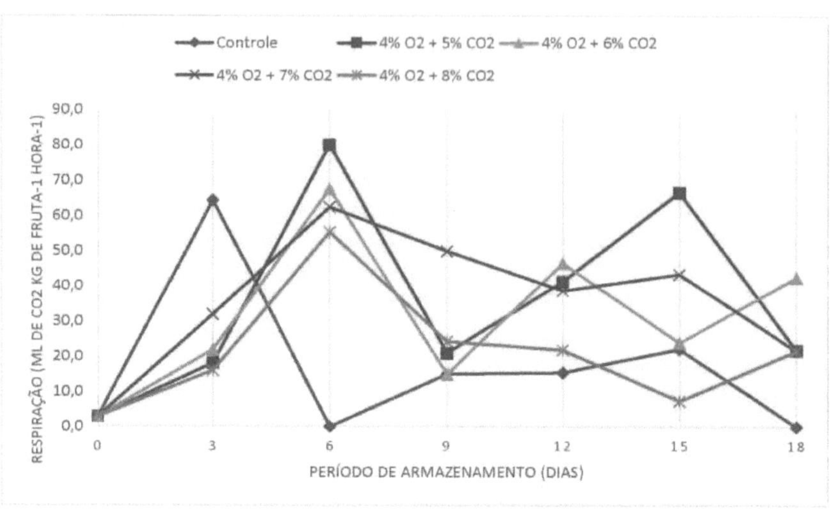

Mass loss in (%)

Figure 44 shows that the treatment with 4% O_2 and 7% CO_2 had the lowest mass loss (0.49%) on the 9th day of storage, 59% less than the control treatment. The lower mass losses (%) obtained with the treatment with 4% O_2 and 7% CO_2 are related to the lower loss of water and solutes during the ripening process, which slows down the ripening process. This statement corroborates Yamashita et al. (2002) who described the behavior of mass loss for fruit in active packaging as being related to an increase in the relative humidity of the air inside the packaging, saturating the environment and reducing the water vapor pressure deficit in relation to the environment, Chitarra and Chitarra (2005) reported that mass losses of up to 10% are acceptable for marketing fruit in general, this statement corroborates Lima et al. (2003), who studied the behavior of graviola and observed that the loss of fresh matter increased during the six days of storage, with the greatest losses being found especially between the second and third day, reaching 4.6%. Reis (2013) obtained weight losses of 2.01% and 3.91% on the 9th day of storage in atemoya cv.'Gefner' fruits packed with PVC film + oxygen absorber; respectively for temperatures of 15°C and 25°C.

Figure 44 - Mass loss (%) in atemoia cv. 'Thompson' characterized by different gas concentrations in active modified atmosphere and stored at 15 ± 0.2°C and 90 ± 2% RH for 18 days. Botucatu SP

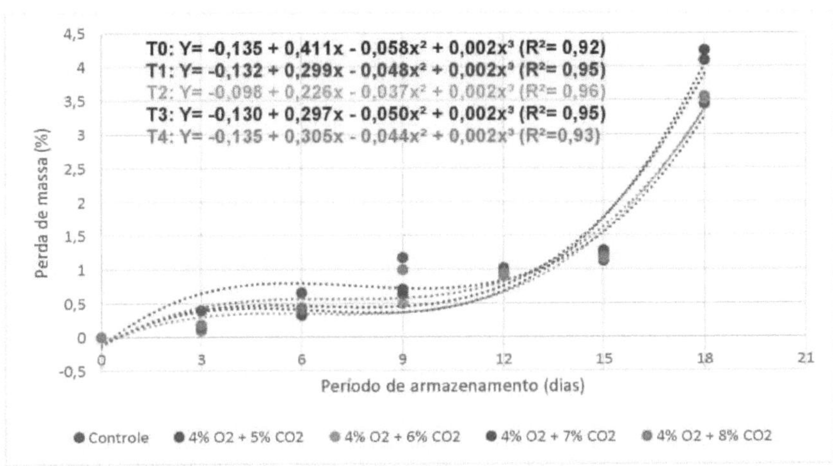

The values obtained for the soluble solids content (°Brix) in the atemoias showed that there was no statistical significance between the treatments and also no interaction between them and the storage period, as shown in (Table 26), which makes it possible to interpret that there was no interference from the different atmospheric concentrations in the ripening process, in terms of the natural ripening process. Figure 45 shows that there was a 136% increase in the soluble solids content between day zero and day 3 of storage, while between day 3 and day 18 there was a 4.7% reduction in the soluble solids content, This can be explained by Chitarra and Chitarra (2005) who report an increase in the concentration of soluble sugars due, for example, to the transformation of starch or the concentration of glucose, fructose and sucrose as the fruit matures, with a subsequent decline due to their use as a source of energy, Lima et al. (2003) observed an increase in soluble solids up to the fourth day after harvest, for graviola fruit stored at 23.4 ± 1.1°C and 81 ± 10.6% RH, and between the fourth and sixth day, the increase was only 0.9 °Brix, so that at the end of the six-day storage period, the ST content was 14.4 °Brix, the data differ from this experiment, possibly due to the different fruit and methodology used, Cia et al. (2010) found values of 11.7% soluble solids for 'Niagara rosada' grapes stored in active modified atmosphere (low density polyethylene - LDPE 50 µm) when they were stored for seven days at 1°C and 90% RH followed by a further three days under ambient conditions at 25°C and 80% RH, Pfaffenbach et al. (2003) found soluble solids levels between 15.4 and 16.0° Brix for mangoes of the early variety 'IAC 103' Espada Vermelha stored in low density polyethylene (LDPE) with a thickness of 25µm and kept in a cold room at 12°C and 90% RH with an ethylene absorber (potassium permanganate), Reis (2013) found soluble solids values of

19,4 and 18.8 °Brix + ethylene absorber, respectively, Santos (2014) observed that jaboticaba fruits stored in partial vacuum with 6μm PVC (low density polyethylene) film for twelve days and refrigerated at 12 ± 1°C and 80 ± 2% RH showed soluble solids values of 14.7%, Sanches et al. (2016) observed that graviola fruits of the 'Morada' variety stored at 23°C and 85% RH, after eight days of storage under pressures of 2.0 kPa O_2 + 2.0 kPa CO_2 and 2.0 kPa O_2 + 3.0 kPa CO_2 showed averages of 15.3 and 15.9 °Brix respectively.

Table 26 - Soluble solids values (°Brix) in 'Thompson' cv. atemoia characterized by different gas concentrations in active modified atmosphere and stored at 15 ± 0.2°C and 90 ± 2% RH for 18 days. Botucatu SP

Treatment	Storage period (days)							Average
	0	3	6	9	12	15	18	
Control*	13,8	18,2	17,4	17,2	19,1	17,0	17,6	17,2
4% O2 + 5% CO_2	13,8	19,3	19,2	20,6	19,3	18,0	17,6	18,3
4% O2 + 6% CO_2	13,8	20,6	18,7	19,7	18,0	18,3	18,1	18,2
4% O2 + 7% CO_2	13,8	16,7	18,7	17,3	18,5	21,1	18,2	17,8
4% O2 + 8% CO_2	13,8	19,3	19,7	19,6	18,8	21,2	18,3	18,7
C.V. (%)	9,5							

The means of treatment and interaction were not significant at 5% probability, according to Tukey's test.

(*) The control treatment is characterized by the fruit being washed in running water and then immersed in a chlorine solution at 150 mg L^{-1} for 10 minutes and then dried at room temperature.

Figure 45 - Soluble solids values (°Brix) in atemoia cv. 'Thompson' characterized by different gas concentrations in active modified atmosphere and stored at 15±0.2°C and 90±2% for 18 days. Botucatu SP

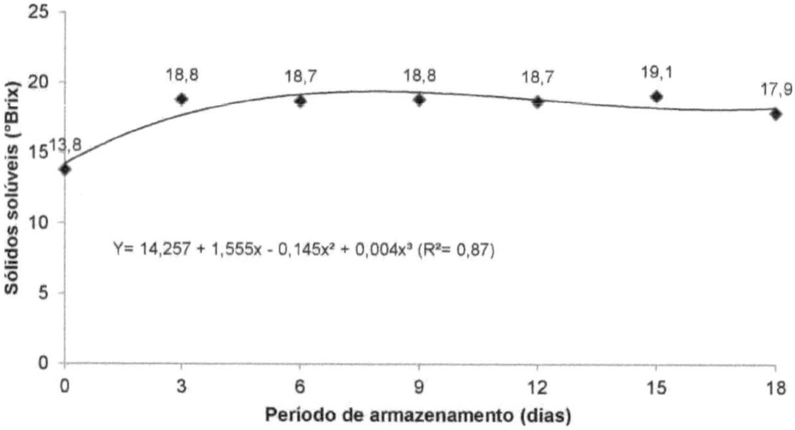

The values found for the pH averages showed that all the treatments differed statistically from the control treatment in terms of their averages over the eighteen days of storage; The analysis of the pH values obtained over the eighteen days of storage showed that there was a gradual increase in pH between day zero and day 18, characterized by a 4% difference. This statement can be explained by the fact that during the fruit ripening process the pH values and soluble solids content increase, the data from this experiment differ from

Sanches et al. (2016) who found pH values in graviola variety 'Morada' after eight days of storage of 3.78 to 3.83 when these were subjected to a modified atmosphere condition of 2.0 kPa O_2 + 2.5 kPa CO_2 and 2.0 kPa O_2 + 3.0 kPa CO_2 stored at 23°C and 85% RH, Reis (2013) found different pH values respectively 4.99%, 5.89%, 4.11% and 2,13% in atemoya fruits stored for fourteen days at 25°C coated with cassava starch and packaged with PVC, PVC + oxygen absorber and PVC + ethylene absorber, this result confirmed that the fruits treated with active packaging containing ethylene absorbers had a better response in inhibiting the ripening of atemoyas stored at 25°C.

Table 27 - pH values in 'Thompson' cv. atemoia characterized by different gas concentrations in active modified atmosphere and stored at 15 ± 0.2°C and 90 ± 2% RH for 18 days. Botucatu SP

Treatment	Storage period (days)							Average
	0	3	6	9	12	15	18	
Control*	5,76	5,88	6,40	6,60	6,55	6,70	6,47	6,34 a
4% O2 + 5% CO_2	5,76	5,78	6,23	6,11	5,94	6,40	5,91	6,02 b
4% O2 + 6% CO_2	5,76	5,89	5,91	6,12	6,11	6,08	6,01	5,98 b
4% O2 + 7% CO_2	5,76	5,80	5,94	6,05	6,09	6,10	5,90	5,95 b
4% O2 + 8% CO_2	5,76	5,80	5,85	6,26	6,14	6,03	6,07	5,99 b
C.V. (%)	3,8							

The means of treatment and interaction were not significant at 5% probability, according to Tukey's test.

(*) The control treatment is characterized by the fruit being washed in running water and then immersed in a chlorine solution at 150 mg L^{-1} for 10 minutes and then dried at room temperature.

The results obtained with the averages of the treatments in the interaction with storage showed that the pH values in the atemoias were characterized by an increase in these values, and in the comparison between day zero and day 18 there was a percentage difference of 4%; This statement makes it possible to interpret that the fruit maintained its ripening process even though it was under active modified atmosphere conditions. This statement corroborates Reis (2013) who observed a slight increase in pH values in atemoyas stored in active packaging, corresponding to a decrease in total titratable acidity values during storage at 25°C for fourteen days.

Figure 46 - pH values in 'Thompson' cv. atemoia characterized by different gas concentrations in active modified atmosphere and stored at 15 ± 0.2°C and 90 ± 2% RH for 18 days. Botucatu SP

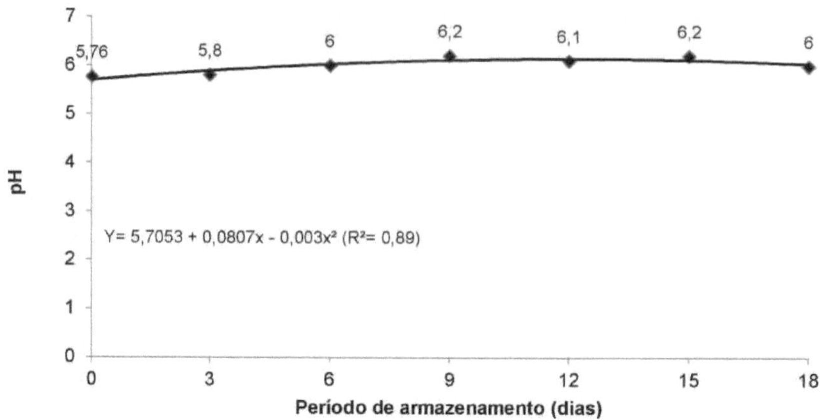

The data obtained on the titratable acidity of the aubergines showed that all the treatments differed statistically from the control, but did not differ from each other or in the interaction with storage (Table 28), which makes it possible to interpret that the use of active modified atmospheres on the aubergines acted positively in maintaining the titratable acidity of the fruit compared to the control, under the same storage conditions, Calegaro, Pezzi and Bender (2002) observed that the use of active modified atmospheres (3 kPa O_2 + 10 kPa CO_2, 5 kPa O_2 + 15 kPa CO_2) for the preservation of strawberries (*Fragaria x ananassa Duch.*) cv. 'Oso Grande' stored in a B.O. chamber.D chamber for (7 and 14 days) at O° C, found that both at 7 days and 14 days, no significant changes in acidity were found, Reis (2013) found in atemoya fruits stored at 25°C in active packaging a lower average reduction in total titratable acidity, compared to fruit treated with edible coatings, with a reduction of 26.83%, 25.20% and 24.12% for fruit packed with PVC and active packaging containing oxygen absorbers and ethylene absorbers, respectively, with the lowest concentration occurring on the 8th day of storage.

Table 28 - Titratable acidity values (g citric acid 100 g^{-1}) in 'Thompson' cv. atemoia, characterized by different gas concentrations in active modified atmosphere and stored at 15 ± 0.2°C and 90 ± 2% RH for 18 days. Botucatu SP

Treatment	Storage period (days)							Average
	0	3	6	9	12	15	18	
Control*	0,137	0,100	0,070	0,063	0,083	0,070	0,073	0,085 b
4% O2 + 5% CO_2	0,137	0,143	0,093	0,103	0,103	0,083	0,116	0,111 a
4% O2 + 6% CO_2	0,137	0,137	0,120	0,090	0,127	0,093	0,113	0,117 a
4% O2 + 7% CO_2	0,137	0,160	0,127	0,107	0,143	0,110	0,100	0,126 a
4% O2 + 8% CO_2	0,137	0,120	0,150	0,083	0,120	0,070	0,087	0,109 a
C.V. (%)	25,5							

The means of the treatment and interaction were not significant at the 5% probability level.

Tukey test.

(*) The control treatment is characterized by the fruit being washed under running water and then immersed in a chlorine solution of 150 mg L^{-1} for 10 minutes and then dried at room temperature.

Figure 47 shows the variation in titratable acidity values during the storage period, and this showed that the reduction was continuous until the 18th day, characterizing a percentage difference between day zero and day 9 of 35%, a percentage that remained high until day 18 of 28.4%, these statements characterize the behavior of atemoia as climacteric, which during their ripening process show a decrease in titratable acidity; The data obtained in this experiment disagrees with Silva et al. (2009) who found titratable acidity values (g citric acid 100 g^{-1}) in 'Gefner' cv. atemoia ranging from: 0.128, 0.085, 0.170, 0.234, 0.234 and 0.213, respectively for days 0, 3, 6, 9, 12 and 15 packed in trays with PVC film, Sanches et al. (2016) found non-significant titratable acidity values (g citric acid 100 g^{-1}) for graviolas stored under different active modified atmosphere treatments (2.0 kPa O_2 + <0.05 kPa CO_2, 2.0 kPa O_2 + <1.0 kPa CO_2, 2.0 kPa O_2 + 2.0 kPa CO_2, 2.0 kPa O_2 + 2.5 kPa CO_2 and 2.0 kPa O_2 + 3.0 kPa CO_2) in a controlled atmosphere for fifteen days and then stored at room temperature for eight days, with values of 0.79, 0.82, 0.81, 0,86 and 0.83 (g of citric acid 100 g^{-1}) respectively, these values differ from the values found in this work, which can be explained by the difference in fruit and treatments, Lundgren (2017) found titratable acidity data in atemoia cv. 'Thompson' treated with different doses of 1-MCP which showed that this increased for the fruit of all treatments up to the 12th day, in line with the pH; these data favor the interpretation of the importance of the study carried out by this experiment.

Figure 47 - Titratable acidity values (g citric acid 100 g^{-1}) in 'Thompson' atemoia cv. characterized by different concentrations of gases in active modified atmosphere and stored at 15 ± 0.2°C and 90 ± 2% RH for 18 days. Botucatu SP

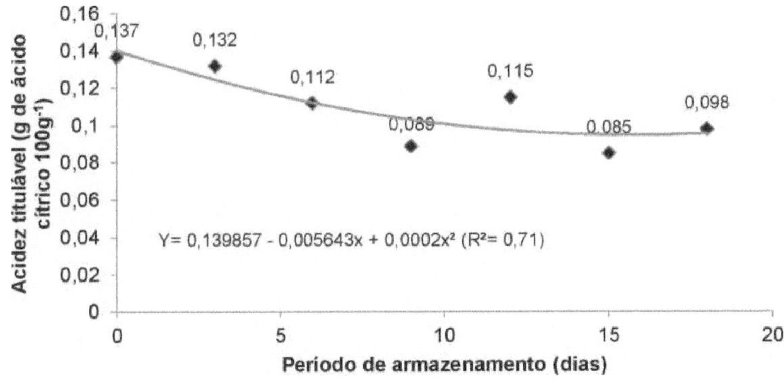

Table 29 shows that the values of ascorbic acid found in the atemoyas were not significant for the treatments or for their interaction with storage (Table 29), which makes it possible to

interpret that the different concentrations of atmosphere did not interfere with the action of the antioxidant (ascorbic acid) during the natural ripening process of the atemoyas.

Table 29 - Ascorbic acid values (mg ascorbic acid $100g^{-1}$) in 'Thompson' atemoia cv. characterized by different concentrations of gases in active modified atmosphere and stored at 15 ± 0.2°C and 90 ± 2% RH for 18 days. Botucatu SP

Treatment	Storage period (days)							Average
	0	3	6	9	12	15	18	
Control*	26,37	20,16	22,82	16,08	19,38	24,26	26,29	22,19
4% O2 + 5% co2	26,37	23,25	14,76	7,47	14,62	18,19	8,09	16,11
4% O2 + 6% co2	26,37	15,46	13,29	19,28	6,93	18,56	8,28	15,45
4% O2 + 7% co2	26,37	6,61	15,19	8,88	10,87	13,37	12,52	13,40
4% O2 + 8% co2	26,37	34,13	26,00	15,19	17,26	17,85	14,48	21,61
C.V. (%)	57,7							

The means of the treatment and interaction were not significant at 5% probability, according to Tukey's test.

(*) The control treatment is characterized by the fruit being washed in running water and then immersed in chlorine solution at 150 mg L^{-1} for 10 minutes and then dried at room temperature.

The data shown in Figure 48 shows a decrease in ascorbic acid values over the course of storage, with the lowest value on the 9th day. This makes it possible to interpret that with the decrease in ascorbic acid values (an antioxidant), the fruit's storage potential decreases, as a result of enzymatic action which can alter the color of the fruit and thus negatively affect the consumer's interpretation of the fruit's quality for consumption.

Figure 48 - Ascorbic acid values (mg ascorbic acid $100g^{-1}$) in 'Thompson' atemoia cv. characterized by different gas concentrations in active modified atmosphere and stored at 15 ± 0.2°C and 90 ± 2% RH for 18 days. Botucatu SP

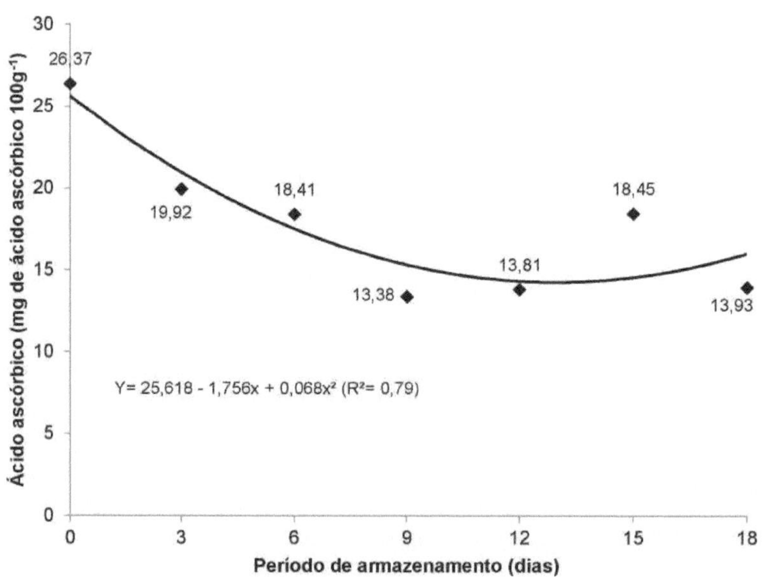

Table 30 shows that the treatments did not differ statistically from the control, nor did they show any interaction with storage or between the concentrations of gases in each treatment, which makes it possible to interpret that the different concentrations of atmosphere did not interfere with the natural ripening process. Figure 49 shows that the average total sugar values (%) were significant over the course of storage, which is characterized by an increase in glucose and fructose during the natural ripening process; This can be explained by Chitarra and Chitarra (2005), who relate the progression of ripening to an increase in the concentration of soluble sugars, such as glucose, fructose and sucrose, with a subsequent decline due to their use as a source of energy.

Table 30 - Reducing sugar values (%) in atemoia cv. 'Thompson' characterized by different concentrations of gases in active modified atmosphere and stored at 15 ± 0.2°C and 90 ± 2% RH for 18 days. Botucatu SP

Treatment	Storage period (days)							Average
	0	3	6	9	12	15	18	
Control*	3,03	4,21	5,96	6,33	5,82	5,98	5,78	5,30
4% O2 + 5% CO2	3,03	4,14	4,85	5,37	4,72	5,40	5,29	4,69
4% O2 + 6% CO2	3,03	3,70	5,37	4,61	4,99	6,23	5,23	4,74
4% O2 + 7% CO2	3,03	3,36	5,24	4,69	5,07	6,48	5,20	4,72
4% O2 + 8% CO2	3,03	4,02	4,70	4,78	4,26	9,05	5,26	5,01
C.V. (%)	21,9							

The means of treatment and interaction were not significant at 5% probability, according to Tukey's test.

(*) The control treatment is characterized by the fruit being washed in running water and then immersed in chlorine solution at 150 mg L^{-1} for 10 minutes and then dried at room temperature.

Figura 49 - Reducing sugar values (%) in atemoia cv. 'Thompson' characterized by different gas concentrations in active modified atmosphere and stored at 15 ± 0.2°C and 90 ± 2% RH for 18 days. Botucatu SP

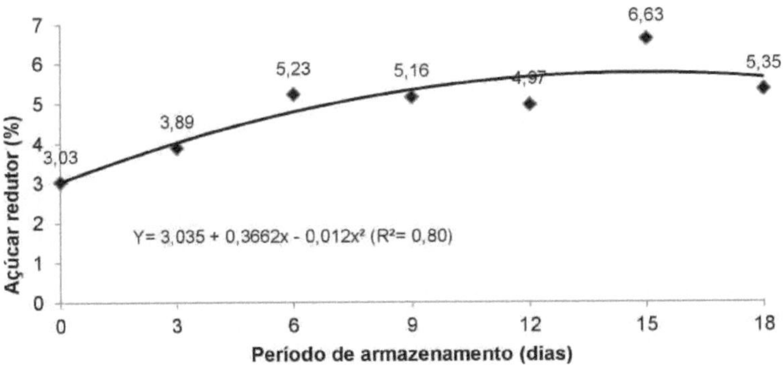

The values found for total sugars (%) in the atemoias showed similar behavior for all the treatments in terms of the evolution of their values over the eighteen days of storage; this statement makes it possible to interpret that the atemoias maintained their respiratory process during storage, with the exception of the treatment with 4% O_2 + 5% CO_2, which showed a drop in its values from the 9th day onwards, This statement can be explained by Chitarra and Chitarra (2005) who report the influence of controlled atmospheres on fruit ripening. In some cases, combinations of gases can be harmful to the fruit, resulting in irregular ripening under conditions of low O_2 levels (< 2%) or high levels of CO_2.2%) and or high levels of CO_2 (>5%), Alves et al. (1997) reported that atemoias have a high concentration of fructose, which is 1.7 times higher than sucrose in terms of sweetness, which means that the sweeter the fruit, the riper it is; Lundgren (2017) found a peak in total sugars on the 9th day of storage in atemoias cv. 'Thompson' treated with different doses of 1-MCP; these statements support this experiment in terms of the importance of quantifying total sugars as an interference factor in maintaining the fruit's shelf life.

Figura 50 - Total sugar values (%) in atemoia cv. 'Thompson', during storage, characterized by different concentrations of gases in active modified atmosphere and stored at 15 ± 0.2°C and 90 ± 2% RH for 18 days. Botucatu SP

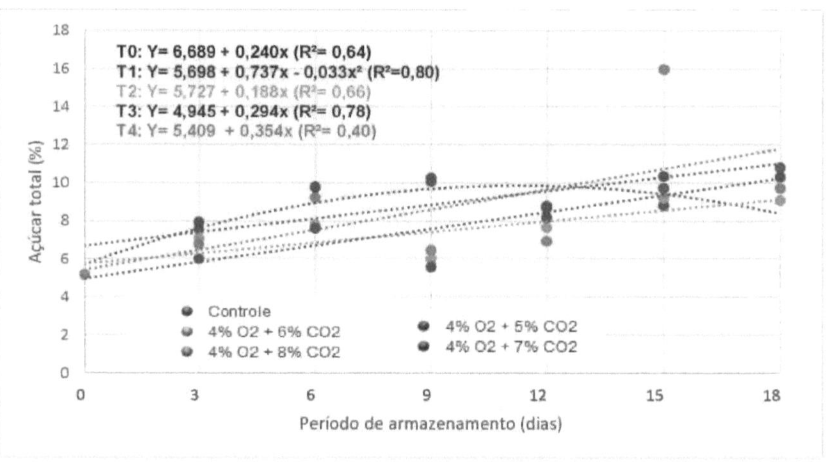

The sucrose values (%) in the atemoyas show that on the 6th day of storage, the treatment with 4% O_2 + 5% CO_2 had higher sucrose values than all the other treatments, which was 18.5% higher than the control treatment; during the same period, the treatment with 4% O_2 + 7% CO_2 had the lowest sucrose value of 2.26%, 37.6% lower than the control treatment; this statement makes it possible to interpret that this treatment delayed the ripening process, due to the decrease in soluble sugars up to the 6th day, these data corroborate with Lundgren (2017) who found the peak for the sucrose content in atemoias cv. Thompson, treated with different doses of 1-MCP on the 6th day of storage (20%), after which sucrose remained similar on days 9 and 12 and decreased at the end of day 18. Chitarra and Chitarra

(2005) explain that the ripening process in fruit is related to an increase in the concentration of soluble sugars, such as glucose, fructose and sucrose, and there may be a subsequent decline in their values due to their use as an energy source. The data from this experiment is better when compared to other experimental methodologies.

Figure 51 - Sucrose values (%) in atemoia cv. 'Thompson' characterized by different gas concentrations in active modified atmosphere and stored at 15 ± 0.2°C and 90 ± 2% RH for 18 days. Botucatu SP

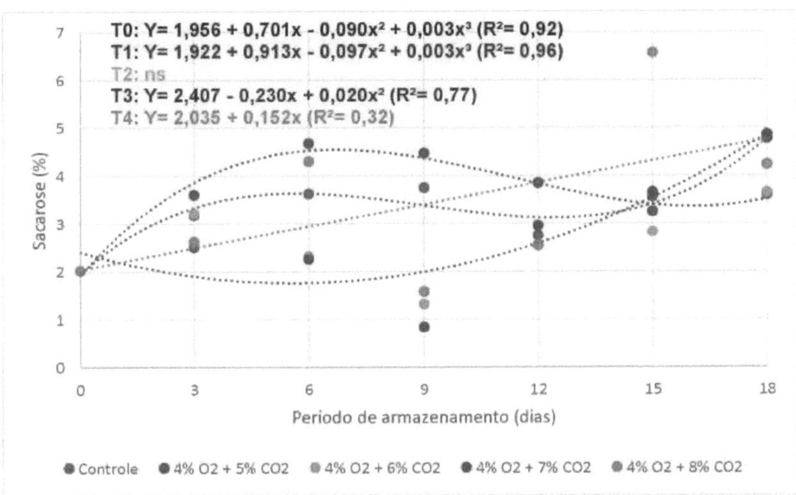

The results obtained for starch (%) in this experiment showed that there was no statistical significance for the treatments, the interaction between them and their interaction with the storage period (Table 31), which makes it possible to interpret the non-interference of the treatments in the alteration of starch concentrations during the natural ripening process of the atemoya.

Table 31 - Starch values (%) in atemoya cv. 'Thompson' characterized by different gas concentrations in active modified atmosphere and stored at 15 ± 0.2°C and 90 ± 2% RH for 18 days. Botucatu SP

Treatment	Storage period (days)							Average
	0	3	6	9	12	15	18	
Control*	18,70	12,37	11,54	14,22	10,14	8,43	11,34	12,39
4% O2 + 5% CO2	18,70	14,32	12,67	13,17	12,98	10,51	7,45	12,83
4% O2 + 6% CO2	18,70	14,50	12,58	17,18	10,72	10,09	9,06	13,26
4% O2 + 7% CO2	18,70	13,41	13,33	14,27	12,58	6,97	9,75	12,71
4% O2 + 8% CO2	18,70	16,71	14,56	15,35	11,59	5,74	9,97	13,23
C.V. (%)	15,3							

The means of treatment and interaction were not significant at 5% probability, according to Tukey's test.

(*) The control treatment is characterized by the fruit being washed under running water and then immersed in a chlorine solution of 150 mg L^{-1} for 10 minutes and then dried at room temperature.

The analysis of the starch values (%) shows in Figure 52 that the starch concentration (%) in the atemoyas compared to the averages obtained between the treatments. Over the course of eighteen days of storage, there was a gradual and constant reduction, with a percentage reduction of 51% between day zero and day 18, This decrease can be explained by the hydrolysis of starch, which is transformed into soluble sugars during the ripening process. This means that the riper the fruit, the lower the starch content and the higher the sugar content, which is characteristic of the respiratory process in fruit.

Figure 52 - Starch values (%) in atemoya cv. 'Thompson' characterized by different concentrations of gases in active modified atmosphere and stored at 15 ± 0.2°C and 90 ± 2% RH for 18 days. Botucatu SP

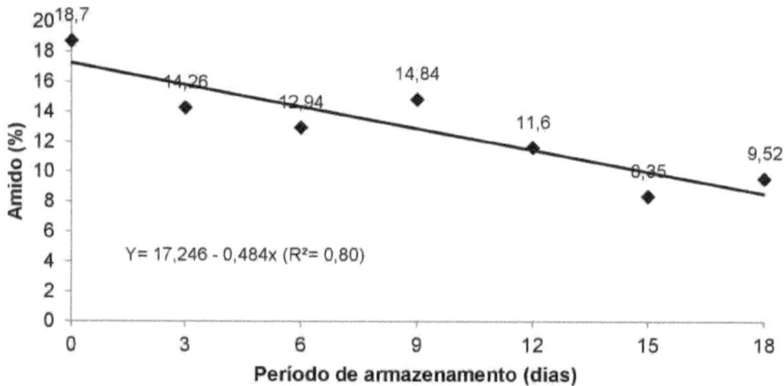

Bark brightness

Table 32 shows that there was no statistical difference between the treatments and the interaction between them. This result makes it possible to interpret that the concentrations of atmosphere tested did not interfere with the natural ripening process.

It is important to note that the data relating to the 18th day of storage has been lost, which explains the table quoting the storage period up to the 15th day; this observation is valid for all the analyses relating to chromatography.

Table 32 - Luminosity values (L*) in 'Thompson' cv. atemoia characterized by different gas concentrations in active modified atmosphere and stored at 15 ± 0.2°C and 90 ± 2% RH for 18 days. Botucatu SP

Treatment	Storage period (days)						Average
	0	3	6	9	12	15	
Control*	56,66	57,82	56,10	48,95	42,17	42,59	50,71
4% O2 + 5% co2	56,66	58,30	57,14	53,44	41,96	46,59	52,35
4% O2 + 6% co2	56,66	56,34	55,72	51,51	42,89	45,69	51,47
4% O2 + 7% co2	56,66	54,44	56,25	48,04	41,89	45,70	50,50
4% O2 + 8% co2	56,66	55,06	57,46	46,85	46,10	44,77	51,15
C.V. (%)	8,1						

The means of treatment and interaction were not significant at 5% probability, according to Tukey's test.

(*) The control treatment is characterized by the fruit being washed in running water and then immersed in a chlorine solution of 150 mg L^{-1} for 10 minutes and then dried at room temperature.

When analyzing the data obtained on the brightness (L*) of the 'Thompson' cv. atemoias, the brightness values ranged from 56.66% on day zero to 43.00% on day 12 of storage. 'Thompson' it was observed in Figure 53 that the luminosity values varied from 56.66% on day zero to 43.00% on day 12 of storage, this being the lowest of the values, this statement makes it possible to interpret that the luminosity showed a grey colour, since L* (luminosity) is characterized on a scale from 0 (totally black) to 100 (totally white); The data obtained in this experiment differ from Reis (2013), who found that atemoias cv. 'Gefner' packed with PVC, active packaging containing oxygen absorbers and with ethylene absorbers, values of 21.76% (PVC), 20.14% (PVC+AO) and 24.81% (PVC+AE), respectively, Lundgren (2017) studying the behavior of skin luminosity in 'Thompson' cv. atemoias treated with different doses of 1-MCP, obtained an increase in the luminosity value until the 15th day of storage, for all treatments. It should be noted that the data from the 18th day was lost, so the relationship of the variation in luminosity (L*) in relation to the averages of the treatments over the course of storage only occurred up to the 15th day. Evaluating the brightness of the skin is an important quality parameter for atemoya, as the skin of these fruits darkens quickly.

Figure 53 - Luminosity values (L*) of the peel in 'Thompson' cv. atemoia characterized by different concentrations of gases in active modified atmosphere and stored at 15 ± 0.2°C and 90 ± 2% RH for 18 days. Botucatu SP

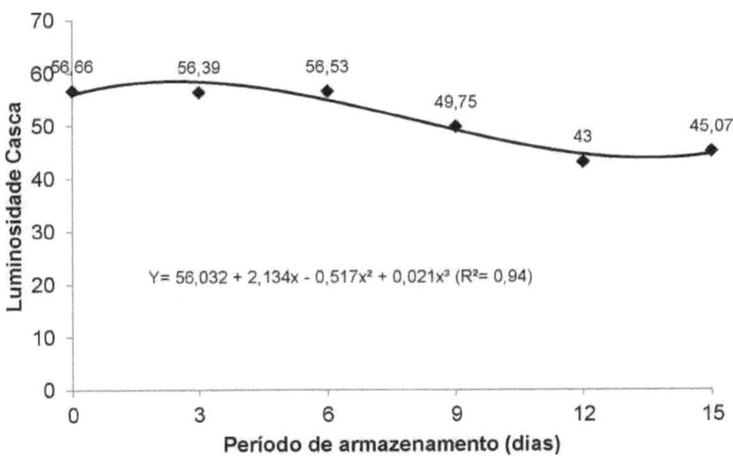

Bark color

Table 33 shows that there was no statistical variation between the treatments and the interaction of the factors studied. This statement makes it possible to interpret that the concentration differences in the atmospheres studied did not interfere in the natural process of variation in the saturation of skin color during ripening.

Table 33 - Shell chroma values of 'Thompson' cv. atemoia characterized by different gas concentrations in active modified atmosphere and stored at 15 ± 0.2°C and 90 ± 2% RH for 18 days. Botucatu SP

Treatment	Storage period (days)						Average
	0	3	6	9	12	15	
Control*	38,24	38,55	39,52	31,43	26,80	24,75	33,21
4% O2 + 5% CO2	38,24	40,12	39,53	38,86	26,93	28,10	35,29
4% O2 + 6% CO2	38,24	39,08	39,10	37,03	26,65	26,82	34,48
4% O2 + 7% CO2	38,24	36,59	39,98	33,21	26,79	26,80	33,60
4% O2 + 8% CO2	38,24	38,41	40,68	33,52	30,91	30,44	35,36
C.V. (%)	10,2						

The means of treatment and interaction were not significant at 5% probability, according to Tukey's test.

(*) The control treatment is characterized by the fruit being washed in running water and then immersed in a chlorine solution of 150 mg L^{-1} for 10 minutes and then dried at room temperature.

The results obtained with the peel chroma values in the 'Thompson' cv. atemoias show that the variation in these values between day zero and day 15 was between 38.24% and 27.27%. 'Thompson' show in Figure 54 that the variation in these values between day zero and day 15 was between 38.24% and 27.38%, which makes it possible to interpret that the chroma of the peel developed a greenish-yellow color, The explanation for the change in color may be related to the degradation of chlorophyll, these data corroborate with Lundgren (2017) who found values for the chroma of the peel decreasing in atemoias cv. Reis (2013) found skin chroma values in 'Gefner' atemoyas of 12.05%, 13.82% and 12.62% for fruit packed with PVC and active packaging containing oxygen absorbers and ethylene absorbers, respectively.

Figure 54 - Peel chroma values of 'Thompson' cv. atemoia characterized by different gas concentrations in active modified atmosphere and stored at 15 ± 0.2°C and 90 ± 2% RH for 18 days. Botucatu SP

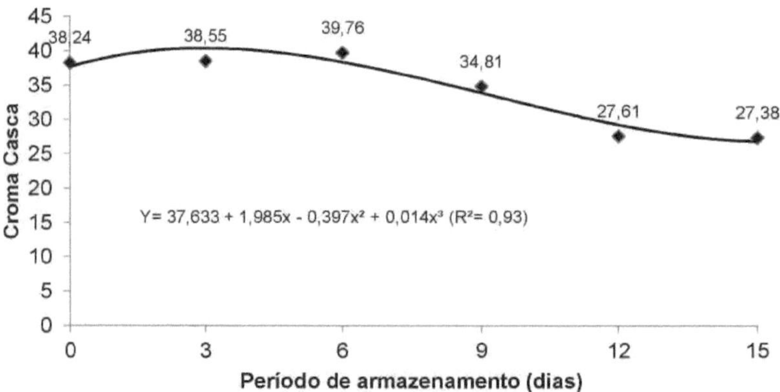

°*Hue* in the Shell

The values found for° Hue in the 'Thompson' cv. tayberries showed that there was no statistically significant difference between the treatments and the interaction of the factors studied (Table 34). This makes it possible to interpret that the variations in the active modified atmosphere used in this experiment did not interfere with the natural ripening process of the atemoya.

Table 34 - °*Hue* values of the peel in 'Thompson' atemoia cv. characterized by different concentrations of gases in active modified atmosphere and stored at 15 ± 0.2°C and 90 ± 2% RH for 18 days. Botucatu SP

Treatment	Storage period (days)						Average
	0	3	6	9	12	15	
Control*	114,2	106,1	106,6	93,1	81,8	77,9	96,62
4% O2 + 5% CO2	114,2	112,1	110,8	105,4	75,2	83,5	100,2
4% O2 + 6% CO2	114,2	111,9	111,6	102,5	77,2	73,3	98,46
4% O2 + 7% CO2	114,2	112,4	111,9	93,5	77,5	74,0	97,26
4% O2 + 8% CO2	114,2	109,9	111,0	95,8	91,7	86,6	101,52
C.V. (%)	8,1						

The means of treatment and interaction were not significant at 5% probability, according to Tukey's test.

(*) The control treatment is characterized by the fruit being washed in running water and then immersed in chlorine solution at 150 mg L^{-1} for 10 minutes and then dried at room temperature.

In the analysis of the data obtained from the° Hue values, Figure 55 shows that the 'Thompson' atemoyas showed an evolution of yellow coloration between day zero and day 15 of storage. These data corroborate Lundgren (2017) who found a statistical difference at 1% probability for the storage time factor of 3, 6, 9, 12, 15 and 18 days for the° Hue of the peel in 'Thompson' atemoyas.

Figura 55 - °*Hue* values of the peel in 'Thompson' cv. atemoia characterized by different concentrations of gases in active modified atmosphere and stored at 15 ± 0.2°C and 90 ± 2% RH for 18 days. Botucatu SP

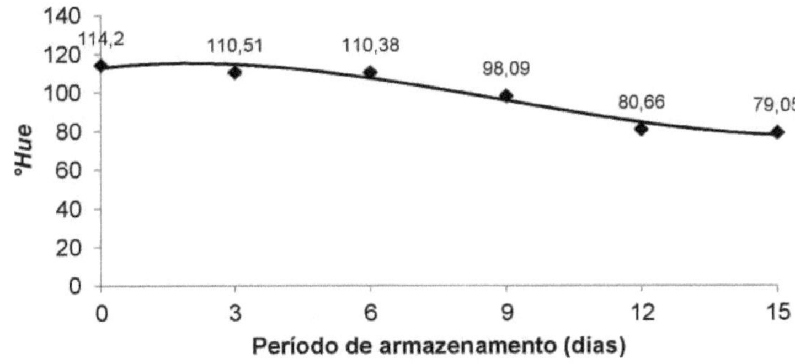

Pulp brightness

The values shown in Table 35 show that the treatments did not differ statistically in terms of pulp brightness in atemoya cv.

'Thompson', as well as in the interaction of the factors studied; which makes it possible to interpret that the different concentrations of atmosphere evaluated did not interfere in the natural process of pulp luminosity during the ripening of the atemoia.

Table 35 - Luminosity values (L*) of the pulp in 'Thompson' atemoia cv. characterized by the different concentrations of gases in active modified atmosphere and stored at 15 ± 0.2°C and 90 ± 2% RH for 18 days. Botucatu SP

Treatment	Storage period (days)						Average
	0	3	6	9	12	15	
Control*	96,01	96,83	95,85	96,48	94,86	95,41	95,91
4% O2 + 5% CO_2	96,01	96,21	95,12	94,99	94,24	95,91	95,41
4% O2 + 6% CO_2	96,01	96,61	95,88	95,30	94,40	95,78	95,66
4% O2 + 7% CO_2	96,01	96,01	95,80	94,20	94,18	94,54	95,12
4% O2 + 8% CO_2	96,01	96,37	95,77	95,37	93,83	92,51	94,98
C.V. (%)	1,2						

The means of treatment and interaction were not significant at 5% probability, according to Tukey's test.

(*) The control treatment is characterized by the fruit being washed in running water and then immersed in a chlorine solution at 150 mg L^{-1} for 10 minutes and then dried at room temperature.

The analysis of the data obtained with the pulp brightness values in 'Thompson' cv. atemoias, shows that the comparison of the averages of the treatments in relation to the storage period, characterized the pulp brightness as white, these data corroborate Reis (2013) who found decreases in the pulp brightness values in 'Gefner' cv. atemoias packed with PVC and active packaging containing oxygen absorbers and absorbers. 'Gefner' packed with PVC and active packaging containing oxygen absorbers and absorbers of 12.84%, 11.32% and 8.08%, respectively over sixteen days of storage and conditioned at 15°C ± 1 and 85% RH.

Figura 56 - Luminosity values (L*) of the pulp in atemoia cv. 'Thompson' characterized by different concentrations of gases in active modified atmosphere and stored at 15 ± 0.2°C and 90 ± 2% RH for 18 days. Botucatu SP

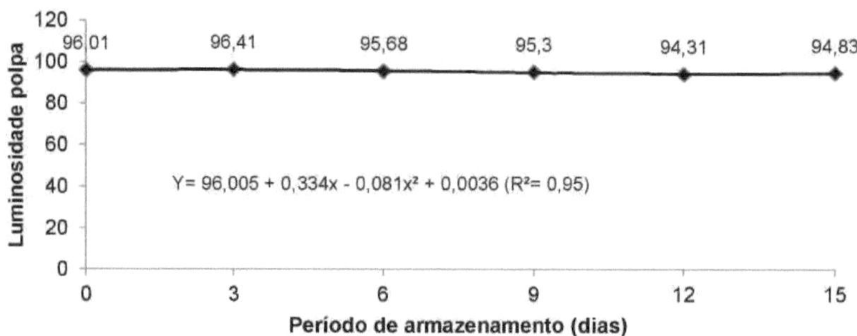

Pulp chroma

The values shown in Table 36 make it possible to interpret that there was no statistical

significance for the pulp chroma values in 'Thompson' atemoyas, for the treatments and interaction factors studied, making it possible to interpret that the concentrations studied did not interfere with the variation in pulp color saturation during the natural ripening process.

Table 36 - Pulp chroma values of atemoia cv. 'Thompson' characterized by different gas concentrations in active modified atmosphere and stored at 15 ± 0.2°C and 90 ± 2% RH for 18 days. Botucatu SP

Treatment	Storage period (days)						Average
	0	3	6	9	12	15	
Control*	10,98	10,59	11,50	10,18	10,58	11,61	10,91
4% O2 + 5% CO_2	10,98	10,15	11,98	10,34	13,37	11,30	11,35
4% O2 + 6% CO_2	10,98	10,33	11,34	10,79	12,57	11,29	11,22
4% O2 + 7% CO_2	10,98	11,02	11,74	11,35	14,26	12,36	11,95
4% O2 + 8% CO_2	10,98	10,62	11,76	11,03	12,12	13,09	11,60
C.V. (%)	10,0						

The means of treatment and interaction were not significant at 5% probability, according to Tukey's test.

(*) The control treatment is characterized by the fruit being washed in running water and then immersed in a chlorine solution of 150 mg L^{-1} for 10 minutes and then dried at room temperature.

The analysis of the pulp chroma values of the 'Thompson' cv. atemoias showed that these values varied in relation to the averages of the treatments and the storage period, characterizing the hue as blue-green, these data corroborate with Lundgren (2017) who found for 'Thompson' cv. atemoias an increase in the pulp chroma values up to the 15th day of storage for all treatments. 'Thompson' treated with different doses of 1-MCP, an increase in pulp chroma values up to the 15th day of storage for all treatments. Reis (2013) found that fruit packed with active packaging (PVC film, PVC + oxygen absorber and PVC + ethylene absorber) were more efficient at preserving the color of the peel and pulp during ripening, compared to fruit from the control batch.

Figure 57 - Pulp chroma values of atemoia cv. 'Thompson' characterized by different gas concentrations in active modified atmosphere and stored at 15 ± 0.2°C and 90 ± 2% RH for 18 days. Botucatu SP

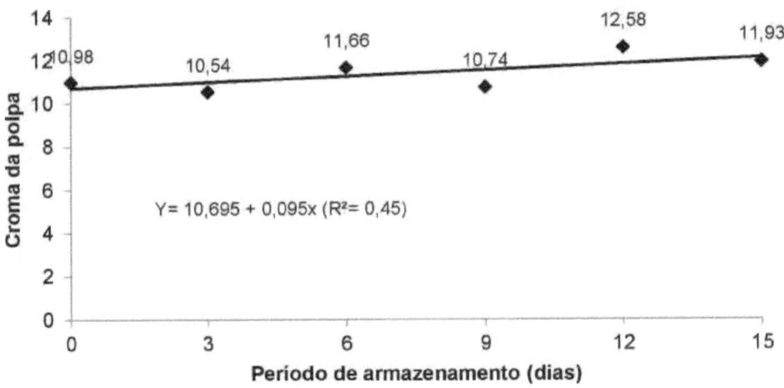

°*Hue* of the pulp

In assessing the ° *Hue* values, Table 37 shows that there was no statistical significance between the treatments or in their interaction with the storage period; this makes it possible to interpret that the active modified atmosphere technology did not influence the variation in the shade of the pulp as it ripened.

Table 37 - °*Hue* values of the pulp in 'Thompson' atemoia cv. characterized by different concentrations of gases in active modified atmosphere and stored at 15 ± 0.2°C and 90 ± 2% RH for 18 days. Botucatu SP

Treatment	Storage period (days)						Average
	0	3	6	9	12	15	
Control*	91,89	92,65	91,89	92,09	91,30	88,99	91,47
4% O2 + 5% CO_2	91,89	91,92	91,28	92,15	90,24	91,76	91,54
4% O2 + 6% CO_2	91,89	91,15	91,40	92,16	91,33	91,80	91,62
4% O2 + 7% CO_2	91,89	91,37	91,99	90,97	89,40	90,39	91,00
4% O2 + 8% CO_2	91,89	91,51	92,96	91,56	91,76	91,00	91,78
Media	91,89	91,71	91,90	91,78	90,80	90,78	91,48
C.V. (%)	1,5						

The means of treatment and interaction were not significant at 5% probability, according to Tukey's test.

(*) The control treatment is characterized by the fruit being washed in running water and then immersed in a chlorine solution of 150 mg L^{-1} for 10 minutes and then dried at room temperature.

Total Antioxidant Activity

The data analysis in Figure 58 shows that the 4% O_2 + 8% CO_2 treatment performed best among all the other treatments during the eighteen days of storage in terms of antioxidant activity in 'Thompson' atemoyas. On the 9th day, the antioxidant activity value was 91.54% and in the control treatment 68.16%, i.e. 134.29% more antioxidant activity over the same period; these data allow us to interpret that the ripening process was slower compared to the control, which allowed for a longer post-harvest shelf life. This statement can be explained by Chitarra and Chitarra (2005) who explain the relationship between an increase in CO_2 gas during storage and a reduction in the product's respiration. The explanation is partly due to the fixation of CO_2 molecules by the malic enzyme (NADPH), which acts by supplying pyruvate for the formation of intermediates in the Krebs cycle. Church (1995) explains the action of CO_2 gas as being responsible for the effect of controlling pathogenic microorganisms. In active modified atmosphere packaging, the indicators that favor its action are related to the amount of gas and its solubility, which is inversely proportional to temperature, so low temperatures have a synergistic effect for the bacteriostatic action (antibiotic action) of CO_2, Blakistone (1999) explains that the action of an active modified atmosphere with 2-5% O_2 and 3-8% CO_2 has been shown to reduce the ripening process of vegetables, with a reduction in softening and microbial action, thus favoring an increase in post-harvest shelf life.

Figure 58 - Total antioxidant activity values (%) in atemoia cv. 'Thompson', during storage, characterized by the different concentrations of gases in active modified atmosphere and stored at 15 ± 0.2°C and 90 ± 2% RH for 18 days. Botucatu SP

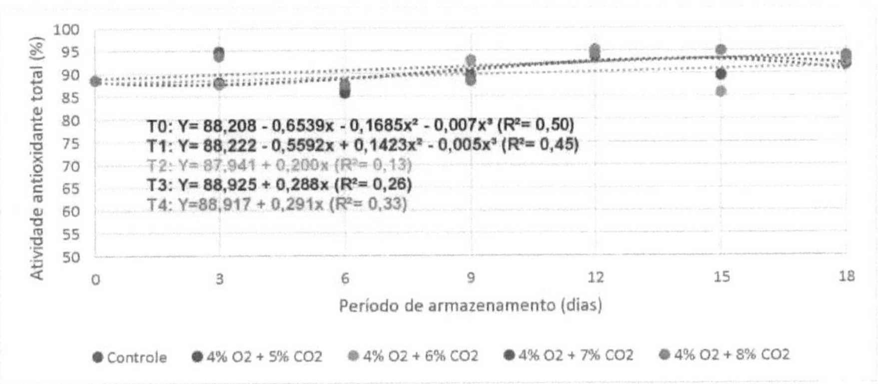

Total phenolic compounds

The data shown in Table 38 demonstrates the lack of statistical significance for the treatments and their interaction with the storage period, which makes it possible to interpret the lack of influence of the active modified atmosphere in altering the values of bioactive compounds during the natural ripening process.

Table 38 - Values of total phenolic compounds (mg of gallic acid $100g^{-1}$) in atemoia cv. 'Thompson' characterized by different concentrations of gases in active modified atmosphere and stored at 15 ± 0.2°C and 90 ± 2% RH for 18 days. Botucatu SP

Treatment	Storage period (days)							Average
	0	3	6	9	12	15	18	
Control *	672,9	829,4	844,7	694,7	1016,6	1823,7	1240,7	1017,5
4% O2 + 5% CO2	672,9	844,5	920,4	514,1	1255,1	1913,8	2842,8	1280,5
4% O2 + 6% CO2	672,9	929,0	952,2	630,6	957,3	1453,5	2544,8	1162,9
4% O2 + 7% CO2	672,9	994,8	833,0	1253,4	1639,2	1943,9	1429,6	1252,4
4% O2 + 8% CO2	672,9	978,7	660,5	928,0	1377,4	1031,8	1752,3	1057,4
C.V. (%)	57,2							

The means of treatment and interaction were not significant at 5% probability, according to Tukey's test.

(*) The control treatment is characterized by the fruit being washed in running water and then immersed in a chlorine solution at 150 mg L^{-1} for 10 minutes and then dried at room temperature.

The data shown in Figure 59 shows that the averages of the treatments between the storage periods were significant for the concentration of total phenolic compounds, with the averages showing that during storage there was an increase in phenolic compounds characterized by 672.9 (mg of gallic acid $100g^{-1}$) on day zero to 1962.05 (mg of gallic acid

100g^{-1}) on day 18, an increase of 291.52%. Phenolic compounds have an antioxidant action that favors the reduction of enzymatic action, such as peroxide, which interferes with the quality of the fruit in post-harvest, in relation to its acceptance by the consumer by negatively altering the color, this statement can be explained by Chittarra; Chitarra (2005) who explain the mechanism of action of the peroxidase enzyme as being based on the formation of enzyme-hydrogen donor complexes, which are found in the cytoplasm of the cell (soluble form) and in the cell wall (insoluble form), making it a potential form of free radicals, which would then act to alter the color internally or externally in the fruit, This statement corroborates Martin-Belloso and Soliva-Fortuny (2006) who report the influence of the types of phenolic compounds and their concentration as direct influencers of enzymatic browning, the data obtained in this experiment differ from Lundgren (2017) who found the highest value of phenolic compounds in atemoya cv. 'Thompson', treated with different doses of 1-MCP, was approximately 950 mg of gallic acid per 100 g of fresh pulp on the 6th day of storage and the lowest was approximately 536 mg of gallic acid per 100 g of fresh pulp on day zero.

Figure 59 - Values of total phenolic compounds (mg of gallic acid 100g^{-1}) in atemoia cv. 'Thompson', during storage, characterized by different concentrations of gases in active modified atmosphere and stored at 15 ± 0.2°C and 90 ± 2% RH for 18 days. Botucatu SP

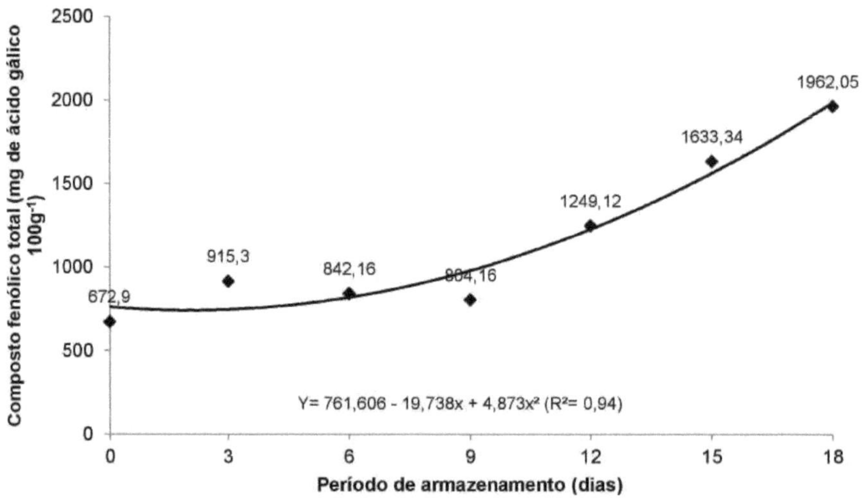

The data shown in Table 39 demonstrates the lack of statistical significance for the treatments, storage and the interaction between these data, which makes it possible to interpret the non-interference of the treatments in the effective control of the polyphenol oxidase enzyme. This information makes it possible to interpret that the active modified atmosphere technology did not reduce the browning caused by PPO in 'Thompson' cv.

atemoia.

Table 39 - Polyphenol oxidase values (μmol catechol transformed min^{-1} g^{-1} fresh matter) in 'Thompson' atemoia cv. characterized by different concentrations of gases in active modified atmosphere and stored at 15 ± 0.2°C and 90 ± 2% RH for 18 days. Botucatu SP

Treatment	Storage period (days)							Average
	0	3	6	9	12	15	18	
Control*	0,404	0,310	0,342	0,279	0,473	0,247	0,250	0,329
4% O2 + 5% CO_2	0,404	0,373	0,374	0,263	0,243	0,401	0,322	0,340
4% O2 + 6% CO_2	0,404	0,289	0,469	0,162	0,331	0,339	0,469	0,352
4% O2 + 7% CO_2	0,404	0,171	0,501	0,233	0,244	0,359	0,503	0,345
4% O2 + 8% CO_2	0,404	0,268	0,325	0,160	0,335	0,296	0,249	0,291
C.V. (%)	50,8							

The means of treatment and interaction were not significant at 5% probability, according to Tukey's test.

(*) The control treatment is characterized by the fruit being washed in running water and then immersed in chlorine solution at 150 mg L^{-1} for 10 minutes and then dried at room temperature.

The data shown in Table 40 demonstrates the lack of statistical significance for the treatments, storage and the interaction between these data, which makes it possible to interpret the lack of influence of the treatments in reducing enzymatic browning during the natural ripening process.

Table 40 - Peroxidase values (μmol catechol transformed min^{-1} g^{-1} fresh matter) in atemoya cv. 'Thompson' characterized by different concentrations of gases in active modified atmosphere and stored at 15 ± 0.2°C and 90 ± 2% RH for 18 days. Botucatu SP

Treatment	Storage period (days)							Average
	0	3	6	9	12	15	18	
Control*	0,042	0,031	0,026	0,040	0,066	0,025	0,034	0,038
4% O2 + 5% CO_2	0,042	0,029	0,040	0,061	0,061	0,043	0,050	0,047
4% O2 + 6% CO_2	0,042	0,066	0,035	0,072	0,067	0,030	0,053	0,052
4% O2 + 7% CO_2	0,042	0,061	0,037	0,074	0,073	0,052	0,033	0,053
4% O2 + 8% CO_2	0,042	0,049	0,042	0,047	0,064	0,046	0,124	0,059
C.V. (%)	55,6							

The means of treatment and interaction were not significant at 5% probability, according to Tukey's test.

(*) The control treatment is characterized by the fruit being washed in running water and then immersed in a chlorine solution at 150 mg L^{-1} for 10 minutes and then dried at room temperature.

3.4 CONCLUSION

The results obtained in this experiment allow us to conclude that:

- The dose of 4% O_2 and 8% CO_2 showed peak respiration in the atemoia on the 6th day, three days longer than the control treatment, which indicates the influence of the active modified atmosphere in delaying the ripening of the atemoia.
- The results obtained with the use of the active modified atmosphere make it possible to interpret that the refrigeration conditions adopted may have been sufficient for the results obtained.

REFERENCES

ALVES, R. E. et al. **Harvesting and post-harvesting of anonaceae.** In: SÂO JOSÉ, A. R. et al. (Eds.). Anonâceas: produção e mercado (pinha, graviola, atemoia and cherimoia). Vitória da Conquista: UESB, p. 240-256, 1997.

AOAC. **Official methods of analysis of the Association of Official Analytical Chemistry International**. 13th ed. Washington, 1015 p., 1992.

BEAUDRY, R. M. Responses of horticultural commodities to low oxygen: limits to the expanded use of modified atmosphere packaging. **HortTechnology**, Alexandria, v. 10, p. 491-500, 2000.

BLAKISTONE, B. A. Principles and applications of modified atmosphere packaging of Foods. New York: Chapman & Hall, 1999. Available at: file:///C:/Users/ghmv/Documents/DOCTORATE/ARTIGOS/MATERIAL%20FOR%20%20ATMOSPHERE%20%20MODIFIED/Atmosphere%20modifi%20each%20in%20conservation%20of%20food.pdf. Visited on 12/02/2018

BLEINROTH, E. W.; ZUCHINI, A. G.; POMPEO, R. µ. Determination of the physical and mechanical characteristics of avocado varieties and their cold storage. **Coletânea ITAL,** Campinas. v. 7, n. 1, p. 29-81, 1976.

BRAZIL. Adolfo Lutz Institute (IAL). Analytical Standards of the Adolfo Lutz Institute -

Chemical and Physical Methods for Food Analysis. 4ª edition, Brasilia: 2008.1018p.

CANO, M. P. *et al* Differences among Spanish and Latin-American banana cultivars: morphological, chemical and sensory characteristics. **Food Chemistry.** n. 59, p. 41119, Amsterdam,1997.

CALEGARO, Jûlio Marques; PEZZI, Ernani; BENDER, Renar Joao. Use of modified atmosphere in the post-harvest preservation of strawberries. **Pesquisa Agropecuâria Brasileira**, Brasilia, v. 8, n. 37, p.1049-1055, aug. 2002.

CARVALHO CRL; MANTOVANI DMB; CARVALHO PRN; MORAES RMM. **Chemical analysis of foods:** technical manual. Campinas: Institute of Food Technology. 121p., 1990.

CHITARRA M. I. F.; CHITARRA A. B. **Post-harvest of fruits and vegetables - Physiology and Handling.** Lavras: UFLA. 785p, 2005.

CHURCH, I. J.; PARSONS, A. L. Modif ed atmosphere packaging technology: a review. Journal of the Science of Food and Agriculture, v. 67, p. 143-152, 1995.

CIA, Patricia et al. Modified atmosphere and refrigeration for post-harvest preservation of 'Niagara Rosada' grapes. **Pesquisa Agropecuària Brasileira**, Brasilia, v. 10, n. 45, p.1058-1065, Oct. 2010.

FERREIRA, D. F. **Sisvar** - System for Analysis of Variance. Version 5. 3. Lavras MG: UFLA, 2010.

KADER, A. A.; ZAGORY, D.; KERBEL, E. L. Modified atmosphere packaging of fruits and vegetables. **Critical Reviews in Food Science and Nutrition**, Boca Raton, v. 28, n. 1, p. 1-30, 1989.

KOBLITZ, Maria Gabriela Bello. **Food Biochemistry.** Rio de Janeiro: Guanabara Koogan Ltda, 2008. 242 p.

LANA, M. M.; FINGER, F.L. **Atmosfera modificada e controlada**: aplicação na conservaçâo de produtos horticolas. Brasilia, DF: Embrapa Hortaliças, 34p., 2000.

LIMA, G. P. P.; BRASIL, O. G.; OLIVEIRA, A. M. de. Polyamines and peroxidase activity in bean (Phaseolus vulgaris L.) grown under saline stress. **Sci. Agricola**, v. 56, n. 1, 1999.

LIMA, M. A. C. de; ALVES, R. E.; FILGUEIRAS, H. A. C.; ENÉAS-FILHO, J. Respiratory behavior and post-harvest quality of graviola (*Annona muricata L.)* 'Morada' at room temperature. **Revista Brasileira de Fruticultura**, Jaboticabal, v. 25, n. 1, p. 49-52, 2003.

LUNDGREN, Giovanna Alencar. **CONSERVATION OF ATEMOIA SUBJECTED TO 1-METHYLCYCLOPROPENE.** 2016. 79 f. Dissertation (Master's Degree) - Agricultural Engineering, Horticulture, Faculty of Agronomic Sciences, Unesp - Botucatu Campus, Botucatu, 2017.

MARTiN-BELLOSO, O.; SOLIVA-FORTUNY, R. Effect of modified atmosphere packaging on the quality of fresh-cut fruits. **Stewart Postharvest**, v. 2, n. 1, feb. 2006. Available at: <http:// www.stewartpostharvest.com>. Accessed on: 12/02/2018.

MENDONÇA, Veridiana Zocoler de. **Physical methods in the preservation of persimmon cv. Kioto in natura and minimally processed.** 2016. 125 f. Tese (Doutorado) - Agronomy Course, Energy in Agriculture, Universidade Estadual Paulista "Jùlio de Mesquita Filho" Faculdade de Ciências Agronômicas Campus de Botucatu, Botucatu, 2016.

MENSOR, L. L.; MENESES, F. S; LEITAO, G. G.; REIS, A. S.; SANTOS, T. C.; COUBE, C. S.; LEITAO, S. G Screening of Brazilian Plant Extracts for Antioxidant Activity by the Use of DPPH Free Radical Method, **Phytotherapy Research**, v. 15, n. 2, p. 127-130, 2001.

MINOLTA, K. **Precise color communication: quality control from perception to instrumentation**. 1998. 59p.

MOSCA, J. L.; LIMA, G. P. P. Respiratory activity of atemoya (*Annona cherimola* Mill. x *Annona squamosa* L.) cv. 'Gefner', during ripening. Proceedings of The **Interamerican Society For Tropical Horticulture**, Fortaleza, v. 47, n. 1, p. 109110, 2003.

NELSON, N. A. A photometric adaptation of somogyi method for determination of glucose. **Journal Biological Chemistry**, Baltimore, v.135, n. 5, p. 375, 1944.

PFAFFENBACH, Luciana Bittencourt et al. EFFECT OF MODIFIED ATMOSPHERE AND REFRIGERATION ON THE POST-HARVEST CONSERVATION OF RED SPRING MANGOES. **Revista Brasileira de Fruticultura**, Jaboticabal, v. 3, n. 25, p.410-413, dez. 2003.

REIS, Maycon Fagundes Teixeira. **INFLUENCE OF MODIFIED ATMOSPHERE AND REFRIGERATION ON THE PHYSICAL, CHEMICAL, TEXTURE AND REOLOGICAL PROPERTIES OF ATEMOIA (*Annona squamosa* L.).** 2013. 147 f. Thesis (Doctorate) - Agricultural Engineering Course, Federal University of Viçosa, Viçosa, 2013.

SALTVEIT, M. E. Is it possible to find and optimal controlled atmosphere? **Postharvest Biology and Technology**, Amsterdam, v. 27, n. 1, p. 3-13, 2003.

SANCHES, Alex Guimaraes et al. Ripening of graviolas after storage in controlled atmosphere. **Revista de Agricultura Neotropical**, Cassilândia, v. 2, n. 3, p.54-61, jun. 2016. Quarterly.

SANTOS, Luana Ferreira dos. **QUALITY AND POST-HARVEST CONSERVATION OF JABUTICABE FRUIT (*Myrciaria cauliflora*) UNDER ACTIVE MODIFIED ATMOSPHERE.** 2014. 61 f. TCC (Graduation) - Agronomy Course, Agricultural Sciences, Federal University of Paraiba, Areia, 2014.

SILVA, Ana Veruska Cruz et al. Use of packaging and refrigeration in the preservation of atemoia. **Ciência e Tecnologia de Alimentos**, Campinas, v. 2, n. 29, p.300-304, jun. 2009. Quarterly.

SINGLETON, V.L.; ORTHOFER, R.; LAMUELA, R.M. Analysis of total phenols and other oxidation substrates and antioxidants by means of Folin-Ciocalteau reagent. Methods of Enzymology, New York, v.299, p.152-178, 1999.

SOMOGYI, M. Determination of blood sugar. **Journal Biologic Chemical**, n. 160, p. 69-73, 1945.

STEFFENS, C. A.; AMARANTE, TALAMINE, C. V. Storage of 'Laetitia' plums in modified atmosphere. **Ciência Rural**, Santa Maria, v. 39, n. 9, p. 24392444, 2009.

VIEITES, R. L.; EVANGELISTA, R. M.; SILVA, C. S.; MARTINS, M. L. Strawberry preservation stored in modified atmosphere. **Semina: Ciências Agràri as**, Londrina, v. 27, n. 2, p. 243-252, 2006. Quarterly.

YAMASHITA, F.; MIGLIORANZA; L.H.S., MIRANDA; L.A.; SOUZA, C.M.A. Effects of packaging and temperature on postharvest of atemoya, Rev. Bras. Frutic., v. 24, n. 3, p. 658-660, 2002.

FINAL CONSIDERATIONS

Considering that Brazil is the third largest fruit producer in the world, behind only India and China, but with post-harvest losses of more than 30% and that the agribusiness sector accounts for approximately 30% of the GDP (Gross Domestic Product), studies into post-harvest preservation are essential to reduce losses and increase the supply of food.

The use of organic acid technology (ascorbic acid and citric acid) to control the enzymatic browning of atemoia demonstrates viability, but its recommended use requires further studies with other organic acids and varying concentrations of those already evaluated.

Ionizing irradiation with cobalt 60 (^{60} CO) shows a positive effect on the respiratory peak of the atemoya when compared to treatment without irradiation, justifying the continuation of the study for doses above 1.0kGy, but following the recommendations of the ABNT (Brazilian Association of Technical Standards).

The study related to the use of active modified atmosphere in atemoia shows results that deserve improvement, both in terms of the study of the atmosphere resulting from the respiration process, as well as the study of other atmospheres starting with 4% O_2 + 5% CO_2, as the best results were obtained from this.

REFERENCES

ANUAR | o BRASILEIRO DE FRUTICULTURA **2017: Brazilian Fruit Year book.** Santa Cruz do Sul: Gazeta Santa Cruz Ltda., v. 1, n. 1,2017. Annual.

BRON, I. U.; JACOMINO, A. P.; APPEZZATO-DA-GLORIA, B. Anatomical and physico-chemical changes associated with refrigerated storage of 'Aurora-l' and 'Dourado-2' peaches. **Pesquisa Agropecuâria Brasileira**, Brasilia, v. 37, n. 10, 2002

CHITARRA M. I. F.; CHITARRA A. B. **Post-harvest of fruits and vegetables - Physiology and Handling.** Lavras: UFLA. 785p, 2005.

CONAB - National Supply Company**. Brazilian program for the modernization of the horticultural market - PROHORT**. Available at: <http://dw.prohort.conab.gov.br/pentaho/Prohort>. Accessed on: March 27, 2017.

JUDD, Walter S. et al. **Plant Systematics:** A Phylogenetic Approach. 3. ed. Porto Alegre: Artmed, 612 p, 2009.

LICHTER, A.; DVIR, O.; ROT, I.; AKERMAN, M.; REGEV, R.; WIESBLUM, A. FALLIK, E.; ZAUBERMAN, G.; FUCHS, Y. Hot water brushing: an alternative method to SO2 fumigation for color retention of litchi fruits**. Postharvest Biol. Technol**, v. 18, n. 3, *235-244*, 2000.

MAIA, G. A., MESQUITA, F. J. A., BARBOSA, M. A. T., FIGUEREDO, R. W. Physical and chemical characteristics of minutes. Pesquisa Agropecuâria Brasileira. Brasilia, v. 21, n. 10, p. 1073-1076, 1986.

MANICA, I. Taxonomy, morphology and anatomy. In: SÂO JOSÉ, A. R.; SOUZA, I. V. B.; MORAIS, O. M.; REBOUÇAS, T. N. H. Anonâceas: produção e mercado. Vitória da Conquista: UESB, 1997. p. 20-35.

MARTINELLI, Gustavo; MORAES, Miguel Avila. **LIVRO VERMELHO DA FLORA DO BRASIL.** Rio de Janeiro: Cncflora, 2013. 1102 p. Available at: <file:///C:/Users/ghmv/Downloads/Red Book (2).pdf>. Accessed on: June 5, 2017.

NEVES, C. S. V. J.; YUHARA, E. N. Characterization of the fruit of atemoya cultivars produced in northern Paraná. **Semina: Ciências Agràrias**, Londrina, v. 24, n. 2, p. 311-314, 2009.

REGO, F. A. O.; ALVES, R. E.; LIMA, E. D. P. A.; SILVA, H.; SILVA, A. Q. Physical and chemical characterization of different fruits from the Annonaceae family. In: CONGRESSO BRASILEIRO FRUTICULTURA, 10., 1989, Fortaleza: Abstracts... Fortaleza: Sociedade

Brasileira de Fruticultura, 1989. p. 493-7.

VIEITES, R. L. Post-harvest preservation of fruit using irradiation. IN: NEVES, L. C. (Org.). **Manual de pós-harvestita da fruticultura brasileira.** Londrina: EDUEL, 2009. p. 189-212.

Printed by Books on Demand GmbH, Norderstedt / Germany